4-9-93 Mardon House 10p.

THE CULT
OF THE EXPERT

by

BRIAN J. FORD

HAMISH HAMILTON

LONDON

First published in Great Britain 1982
by Hamish Hamilton Ltd
Garden House 57–59 Long Acre London WC2E 9JZ

Copyright © 1982 by Brian J. Ford

British Library Cataloguing in Publication Data
Ford, Brian J.
    The cult of the expert.
    I.   Title
    828′.91049   PR6056.0/
    ISBN 0–241–10476–9

Filmset by Saildean Phototypesetters Ltd
Printed and bound in Great Britain by
Richard Clay (The Chaucer Press) Ltd, Bugay, Suffolk

Contents

Introduction

No-one who reads this book has escaped from the effects of the cult of the Expert. Whether it is food or drug standards, the direction of traffic down a favourite street, what you hear on the radio, see on TV, eat, drink, or die from, there is an Expert at work. New ideas sweep the board, only to vanish in a puff of mediocrity when fashions alter. And all the while, incalculable sums of much-needed finance evaporate in the sustenance of glossier and ever more demanding research projects in which individuals with small minds embrace ever more ambitious revolutions.

No matter how hungry the world becomes, or how hard-pressed the economy, the Experts maintain their hold on the power-structure of society. Yet if we are to make sense of the future, it is time the phenomenon was deglamourized and put permanently into a clearer perspective.

The great achievement of Experts has been the dramatic change in Western society. A few decades ago people emerged from school and college literate and well-educated. That has now been changed. Today's young adults are less literate, less educated, and more demoralized than any generation in living memory. It is vitally important that society continues to take this downward path, from the Expert's point of view. This is the surest way of keeping the public in their place.

But it is important that the effects of the deterioration are disguised. Experts have evolved two methods of doing this. The first is the progressive lowering of standards, so that people gain spurious qualifications. The kind of pupil who would have failed school certificate examinations a generation ago is now guaranteed a good pass. This is partly because the

7

pupils put pressure on their teachers to give them satisfactory grades, or they will never find a job. It is also because teachers know that successive bad examination results reflect adversely on their performance as teachers – so they make sure that almost everyone passes, with self-interest in mind.

At the other end of the spectrum, the job-description business has continuously up-graded vacancies so that they sound progressively more prestigious as time goes by. The staff in laboratories who were known as laboratory assistants in the 1940s had been re-named technicians by the 1950s. They became experimental technologists for the 1960s and finally became laboratory scientists by the 1970s.

Yet at the same time, the complexity of their jobs was steadily dropping. Culture media and chemical reagents that used to be made up with care and understanding are now ladled from disposable containers. Readings from dials are all automated, so that the amount of brain-power used in obtaining results is minimal. The way things are moving I would not be surprised to find all these people re-designated hyperscientific professorial fellows by the time the decade is out, for by then all they will have to do is switch everything ON in the morning and OFF at home-time.

It is easy to perceive the roots of the anti-science movement which has flourished (most notably in the United States) in recent years. So bitter has been this back-lash that some commentators are now fearful for the very future of scientific research. In my view the confrontation between pro-science and anti-science factions is wide of the mark. No matter how you define it, the concept of 'science' implies the gaining of natural knowledge for the benefit of mankind. The only alternative would be the anarchy of ignorance. To be anti-science is an absurdity.

It is not science that is the enemy. The problems arise when the superficial manifestations of science are exploited by the materialistic, the acquisitive, the pompously ambitious. At this stage, 'sciencism' becomes an industry in its own right. If you have the right long words, no matter if there is no meaning behind them. If you have the grandiose aim, no matter if there is no integrity. If you are obsessed with power, no matter if you lack wisdom or worldliness. You have now become an Expert, and your craft is the converse of altruistic science. It is Nonscience.

The mushrooming growth-industry of Nonscience (pronounce it how you will) is the real enemy. Its monstrous power structure is nothing more than a confidence trick. The public, who are too poorly educated these days to tell the difference, have been encouraged to accept a range of beliefs that all serve to keep Experts in their unchallengable position. At the top of the list is the notion that the best brains are the most specialized. This is nonsense. The greatest innovations in the development of society, and certainly in the history of science and technology, have been made by inspired individuals, and not by indoctrinated specialists.

They ranged from the violinists who perfected colour photography and the mortician who developed automatic telephone dialling to the draper who discovered bacteria and the schoolmaster who founded space travel. Specialization has never been the hallmark of greatness. Our belief that it is has almost eliminated people with a broad, all-embracing, relevant education. I have no doubt that we should aim at 'holism' (a word coined back in 1926 by General Smuts), where our sages for the future know about a host of facets of the world, and try constructively to understand the breadth of human endeavour. Today's era of specialization has done much to threaten our survival, through short-sighted planning that has left us with a burden of pollution future generations will resent.

As it is, we are increasingly ruled by men and women who know little about anything apart from their own capacity for power. The very people who have the greatest abilility to threaten us all are uniquely narrow-minded, and almost culturally illiterate.

9

We do not need a race of moronic super-specialized Experts who know nothing about almost everything, but wise and insightful scientists who try to understand what little we know so far.

The contorted language of the Expert, designed to keep outsiders at bay rather than to communicate, is a hallmark of Nonscience. Experts do not use ordinary tools like scissors, scales, or spirit-levels. Instead they demand digital-readout electronic laminate cutters, video-display compatible mass-indicators, or gravity-responsive inertial laser modules, in terms outsiders would scarcely comprehend.

Yet their language embodies a heady illiteracy. Experts all talk of the 'media' as a singular noun, so that you hear them speak of 'television as a powerful media' in a discussion on 'medias' in general. Experts think that 'minuscule' is spelt 'miniscule'. They build verbs out of nouns like 'to haemorrhage' and 'to interface', and then use these terms ungrammatically (so that they speak of equipment that can 'interface to' computers, rather that 'interface with'). I have just seen a description of an experiment in which the liquid in some sample tubes split into two phases (like oil floating on water, for example). Said the description alongside: 'Note the tube that didn't two-phase'. Note also the reader who didn't appreciation ...

People who go to conferences are no longer called 'attenders'. They are now known instead as 'attendees', as a result of this corporate illiteracy. That is no more logical than using the term 'prison escapees' for the warders. Conferences themselves are not what they were. The papers are circulated beforehand, so that nothing new results from going to the lectures, and the participants have plenty of time instead to blind any nearby reporters with enough high-flown obscurantism to ensure their reputation remains unsullied.

The simplest term can be instinctively be made complex by the Expert mind. One elementary means is the *rule of superloquation*: 'never use one word where three will do'. Instead of 'head', write CAPITATE

SENSORY EXTREMITY. Rather than 'car', say MOBILITY TRANSPORTATION SYSTEM. Do not say 'book', but LITERARY DATA -ACQUISITION FACILITY. You can continue the process indefinitely, by replacing each of those Expert terms with three others (capitate = MODULAR CEREBRALISATIONARY FUNCTIONAL, sensory = NEUROCHEMICAL DETECTION-ORIENTATED SYSTEMIC, extremity = DISTAL PROJECTIONALIZED CONFIGURATION) so that the simple word 'head' at once becomes a 'modular cerebralizationary functional neurochemical detection-orientated systemic distal projectionalized configuration' with the minimum of effort. Some Experts have been known to take as long as 20 minutes to ask what time a train is, only to find it has left by then.

This same emphasis on words or data, rather than on ideas, is also short-sighted. A large research grant was recently consumed by a hormone specialist who was looking for prostaglandins in menstrual blood (prostaglandins have been highly fashionable in recent years). He spent years identifying his new discovery, by analysing sanitary towels which were saved in plastic bags by all his female staff. Only after his findings were published did it turn out that all he had detected was a chemical constituent of sanitary towels, and nothing to do with human hormones at all.

Once a viewpoint is fashionable, there is no end to the slavish support it attracts. Millions of pounds can be diverted into repetitive investigations by innumerable institutes, all chasing the same trendy topic whilst other, equally deserving subjects go by default. Abortion is one hot topic. For decades the idea was anathema, and the very concept of termination of pregnancy was tantamount to murder. When the fashion dramatically changed in the 1970s, anyone who suggested abortion might be a bad thing was immediately jumped on. A television playwright who was asked to prepare a major dramatic opus for late 1981 tells me that, when she told the producer it would be a play about the ethics of abortion, he

11

replied: 'It will be strongly *pro*-abortion, of course?'

'Does that matter?' retorted the writer.

'Of *course* not, dear,' came the soothing reply. 'You may write what you wish. But if by any chance it *isn't* pro-abortion, well, you know how fashionable the whole thing is right now and it might just not get a screening, with pressure being so tight ... ' I had not been told that more than a couple of weeks when there was a near-riot at a liberal street carnival in the local town. One of the stalls was going to be manned by anti-abortion protesters, and all the other stall-holders signed a petition announcing that, if there was any of that kind of talk, then they would all pull out. Nobody is concerned with the ethics of being pro- or anti-, mind; it is a question of whether you follow the fashionable line or not. Stepping out of fashion these days is a serious offence.

People will say anything to score a victory. We all know that margarine has been much-publicized as being 'better than butter'. Since margarine is a fatty solid, just like butter in that sense, it is just as fattening as butter. So there is little justification for regarding it as somehow miraculously better for us. Recently the British butter authorities decided to hit back with a marketing campaign of their own. They published large advertisements proclaiming that margarine was made of all manner of unsavoury-sounding substances, like tallow and oil, whilst butter is nothing but cream and a little salt.

MARGARINE COULD BE a major killer—and the working class may be the main victims, a reader in chemistry at the Polytechnic of Wales claimed today.

As it happens, all normal butter contains chemical colouring. The advertisements were nothing more than a downright misrepresentation. This area is an excellent example of the way that public opinion is moulded by Experts who make proclamations that ordinary people are in no position to challenge.

A contentious health issue is whooping-cough vaccination. 'How would you like it if your child ended

12

up severely ill, or even DEAD?' said a vaccination campaigner the other day. He was trying to emphasize the need for whooping-cough vaccine, but the reasons he quoted are exactly those that make many parents avoid this form of immunization in the first place. In fact, vaccination against whooping-cough is not the panacea it appears. The death rate from the disease had already fallen to a low level in the late 1950s, before vaccination was introduced. There was no further fall throughout the years of immunization until 1977-79. But by then the level of children receiving the vaccine had already fallen below 50 in some areas. Many children who had not been vaccinated escaped the disease, whereas of those who did become ill with the disease about 30% had been immunized previously. Meanwhile there is a residual hazard of brain damage in a small proportion of the treated children. These matters ought to be put sensibly in perspective; the one-sided propaganda that is all we ever hear does justice to nobody, least of all the children.

In just the same way, we like to preach that antibiotics have helped us banish diseases. Though there is obvious truth in it, we should also realize that the level of most infectious bacterial diseases had fallen dramatically long before antibiotics were available, largely because of better hygiene and public health. What is more, there is still no readily-available cure for any of the virus diseases (from coughs and colds to lassa fever or vervet monkey disease) and few Experts boast about that.

Meanwhile there is a rising incidence of scurvy amongst the elderly. This mediaeval condition is re-emerging as the junk-food era fills more people with a deficient dietary intake. Amongst the young there is an increasing amount of infestation with head lice. There are already signs that life expectancy in the developed countries is going to fall, rather than rise; and the epidemic of cardiac disease (now the Number One killer) remains uncontrollable. If you have a heart attack, there are plenty of spokesmen

13

who will tell you to take as much exercise as you can, to build up those damaged cardiac muscles. There are just as many who will advise you to rest as much as possible, to give the muscles a chance to recover. This principle of contradiction is found everywhere. Half of the world's prognosticators will warn of the burning of fuels, and the increase of carbon dioxide, which will cause a dramatic over-heating of our planet through the 'greenhouse effect'. The other half will point out that the increasing smoke levels that also result are cutting off the amount of solar energy we receive each day, which will precipitate a new ice-age any time now.

Meteoroligists may not have succeeded in 'controlling' the weather, but every corner of the globe has been infiltrated by the products of the pesticide industry, and contaminated by the fallout residues of the nuclear weapon Experts. It is in the military field that the greatest expenditure is found, and where the most muddle-headed thinking runs rampant. Few people realise that the casings and gear-box housings of modern vehicles of war – notably planes, helicopters and warships – are made of aluminium or magnesium-aluminium alloys which catch fire when hit. As the Falkland Islands crisis showed in 1982, modern warships can burn like fireworks. Only the deep-seated ignorance of Expert thinking could make such a fiasco possible.

The fact is that, for all the veneer of sophistication about our hi-tech modernity, we know astonishingly little about the whole spectrum of science. But no Expert ever admits that. It is a cardinal rule of the Expert code that you always simulate total confidence. I have in front of me as I write a book on tuberculosis which states that the disease 'is no longer the scourge it was once', and can now be successfully cured. The 'cure' consisted of burning the skin with a blistering mixture of iodine and alcohol, for this book was written in the mid 1800s. There was no mention of the fact that the cause of tuberculosis was unknown, that there was very little known about treatment, and that to speak of a 'cure' was a calumny at the time. Today's works on diet, fitness, psychology, sociology and a host of other specialist areas are just as overconfident.

To overcome these appalling deficiencies, we would need to do several things:

FIRST, to begin to analyse what I have called the 'areas of ignorance' that surround us, so that research priorities could be identified.

SECOND, to aid research between disciplines, and to aim towards complete, rounded, generalist education.

THIRD, to discourage profligacy, and to end duplication of effort; the end being to fit research for the demands of an over-stressed world.

For the present, that is a personal pipe-dream. Meanwhile – as a prophylactic for the immediate future – we should come to grips with the way Experts work, and how Nonscience functions. That is the most reliable way to keep a hold on our sanity.

Which explains why I have written this book.

15

CHAPTER ONE

Enter the Expert

We have scarcely noticed the fact. But for the major issues of our lives democracy is dead. A new race of super-motivated Experts has silently slid into power and, between them, have taken over the major decision-making processes in this teetering society of ours.

Experts are unassailable and superior individuals who use a language of their own to cloak their inner whims in a spurious aura of authority. As a result, everyone from the man-in-the-street right down to the politician takes very seriously what they say. All Experts delight in changing the world we know and love: not necessarily for the better, perhaps even for the hell of it.

From cyclamates to cholesterol, from the classification of schools to the controversy over Concorde, the work of Experts is all around us and nobody can be sure what next is going to be introduced, altered or outlawed. All we can guarantee is that Experts will go on taking decisions that alter our lives and, whether it is birth, survival or doom, you can rest assured that an Expert will be along any moment to change some aspect of it when you least expect it.

The power of Experts is hard to imagine, but it approaches that of the emperors of old, and compares favourably with the ancient gods of folklore. In earlier centuries, mystics or village sages, with spells and muttered incantations, used secret charms to mystify the neighbours and subdue the opposition. The language of today's Expert is directly evolved from this, and his long words act as tokens which provide that cloak of magic and mystery. There is an equally interesting comparison between the coloured lights and hypnotic whirr of laboratory apparatus and the

17

frightening spells cast by witch-doctors in the tribal village communities. It is all the same, really, and our touching faith in Experts is nothing more than a direct descendant of pagan worship. That is why we are taking Experts so seriously, just at the time when organized traditional religious values are so notice-ably on the wane.

As the old religions die out, new ones take over. The juxtaposition of today's esoteric research with yesterday's symbols reminds us that faith goes on, even when the target for it alters.

There are names for this special Expert language, of course; it is variously known as gobbledeygook, expertese or obscurantism. Its particular purpose is to make sure that all outsiders are kept at bay, and to enable one Expert to recognize another in a crowd. It is a fundamental mistake to imagine that the words are there to aid communication; their true purpose is exactly the converse.

An additional justification for this mysterious tongue which Experts like to speak is that it helps avoid any come-back when things go wrong. A term like *theoretically viable*, for example, can be used to make the most dastardly proposition look respectable at the time of its introduction, but is just as useful to escape from blame if it all goes wrong *(theoretically viable, though not practically feasible*, it then becomes).

Take the well-worn word *situation*. The old-fashioned person would say that his car had broken down, but the modern Expert has introduced the mechanical breakdown situation, which automatically acquires a more authoritative technical and somehow esoteric implication. Today's young people might talk about 'an ongoing experience situation at this moment in time' instead of a bald 'experience'; and one notable member of the United States legislature has already begun to replace 'at this moment

In the past year I have found myself in two consulting situations where management said it wanted to operate in a participative decision-making mode. That is, management said it wanted to share information with the people to be directly affected by the decision, and give them real input into the decision-making process. One was in a labour relations situation at a television station. The other was in a university marketing situation where diversification into a new area of education was being considered. In both situations there arose the issue of how inclusive the management consulting report should be that was to be shared with many different people not in the traditional management decision-making line.

Richard P. Nielsen, Personal Review

Reprinted by permission of *Private Eye*

in time' with *'at this juncture of maturation in the situation'* which is beginning to approach the state of meaninglessness contortion that every Expert longs for.

Do not imagine that the secret language of the Expert is solely intended to keep out the ordinary public. Such words and phrases are used to keep anyone else out – including other Experts and, even more important, the members of the committees whose decisions fund Experts and their activities. A few choice paragraphs of elongated and contorted pseudo-prose can keep the most dazzling mind at bay. Experts know that the right words can change a mundane and pointless programme of research into a world-beating exciting campaign on the very verge of an epoch-making breakthrough, and that that is how to attract funds.

No matter how gloomy are the economic forecasts, or how over-stretched the resources of the state, Experts keep getting money (usually in steadily increasing amounts) whilst the rest of us go bank-rupt. Commercial companies, tradespeople, business organizations, shops, stores, cab-drivers and milk-men need to be financially successful if they are to keep in operation. But Experts have a unique form of financial support, called 'grants', and these have the propensity of being self-perpetuating and often steadily increasing in size and scope as time goes by.

If you were a businessman whose design for a new kind of garden weeder was in trouble, you might well find that it was impossible to obtain further finance without some sound evidence that justified it. 'If your garden weeder doesn't weed,' you would be told, 'forget it.' No Expert expects to have that kind of reception. For him, the future of his *species-specific termination device* (especially if it has possible applications for Third World utilization and uses renewable energy sources) is assured. Not only would he get lavish financial support for the work, but he might even be offered a trip to listen to someone else talk

20

about weeding in some far-off resort, all expenses paid and plenty of opportunity to take his family too for a week's holiday as part of the deal.

What is the difference between the businessman and the Expert? First, not only is failure more acceptable in the Expert world than in the world of commerce, but it can be a positive advantage. Only if further grant aid is forthcoming, runs the tale, can the long-awaited breakthrough materialize. With care and with the right applications of the Expert's special knowledge, this can provide a life-time of financial security and prestige for the very minimum of actual achievement. The Expert never goes 'cap in hand'. His request for funds is couched in arcane language using unintelligible expressions. He commands awe and always manages to convince a grants committee that he knows more than any of them so he must be right.

Further, Experts claim rights which are denied to most businessmen. These days it is hard to fire an employee at whim. Short of video-recording him in the act of theft or violence against a respected senior member of staff, sacking someone is about as hard as inducing them to change sex. Experts use the grant system to overcome this tedious interruption to life's otherwise well-governed tenor. Whilst acknowledging that a member of their staff cannot arbitrarily be sacked they can naturally arrange for their grant support to be redirected. In this way, an Expert can remove the livelihood from an unusually rebellious or undesirable member of staff by the efficacious means of cancelling their income.

Experts are above ordinary law-enforcement agencies. Take the research carried out by Stanley Milgram at Yale in the early 1960s. He gave instructions to volunteers which were clearly unlawful. At worst, his commands (so far as the volunteers were concerned) could have killed someone. But the number of volunteers who questioned the wisdom of what they were told to do was very small indeed. This is an example of the way that the public (who are always

21

ready to question a local official, a policeman or a caller at the door) will do just what an Expert says, no matter how unreasonable, simply because the command is being given by an Expert and not by an ordinary member of society. Not only are Experts above the law; they are equally untouched by the moral code.

Experts as Gods

Where did Experts come from? The alchemists who toiled towards the philosopher's stone that would turn all base metal into gold had none of the unassailable prestige of the modern Expert. If there are antecedents for us to study, we must look away from the Expert's present-day environment of the lavish institute and concentrate instead on the god-like qualities of power that all Experts enjoy.

It is here, in the pursuit of power, that Experts have their soul-mates from earlier ages of mankind. In the tribal races you may find the witch-doctor, striking fear into everyone he chose to threaten (including the rulers of the society) and carefully maintaining an unassailable superiority. How did he do it? There were several distinct means:

i: by using characteristic dress, different from that of the ordinary people;
ii: by carrying strange and mysterious symbols of authority such as tokens, charms or magical wands;
iii: above all, by speaking in a haunting and mysterious language, rich with complicated and intimidating words and unfamiliar phrases, which anyone but a witch-doctor could never interpret.

For the Expert the comparison is clear.

a) His dress is certainly characteristic, for Experts have borrowed from their scientific counterparts the distinctive white coat of authority which sets them above ordinary folk. It is worth noting that Dr Milgram's research suggested that it was the white

22

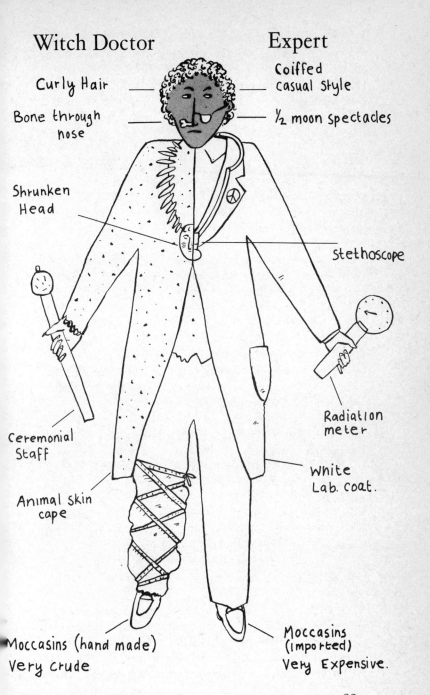

Witch Doctor

Curly Hair

Bone through nose

Shrunken Head

Ceremonial Staff

Animal Skin cape

Moccasins (hand made) Very crude

Expert

Coiffed casual style

½ moon spectacles

Stethoscope

Radiation meter

White Lab. coat.

Moccasins (imported) Very Expensive.

23

coat of the Experts which went far to impress the volunteers that they ought to do just what they were told.

b) Experts carry symbols and charms, too. A stethoscope dangling round the neck (even when the wearer is far from the environment where he might ever actually use such a thing) is one example. Talking or printing calculators are another. A thermometer or a pair of scissors, perhaps the handles of artery forceps, or some kind of special light in a top pocket are what Experts utilize to obtain the same kind of instant recognition as a witch-doctor.

c) Today's laboratories (like the dimly lit den of yesteryear) are cluttered with sleek, complex items of equipment. They are there to impress. Visitors are told how much such devices cost, how powerful they are, and how they could easily provide that breakthrough on which everything depends, from a cure for cancer or the common cold to the origins of life or the control of anything from the weather to the human mind.

What outsiders are rarely told (since there is little in this area to tell) is what the machines have actually done. The point is what they are going to do one day, or what they could do in theory. Anyone who looks for startling past achievements from these costly items of capital investment is going to be disappointed. Experts acquire new apparatus. They rarely have any idea how to use it.

d) And the haunting and mysterious incantations are well to the fore in all Experts' speech. Whereas it might once have been 'abracadabra' that mystified the public, it is now *transitional correlative contingency* or *reciprocal soft-ware strictures*.

Two-word adjectival constructions are a mainstay of the Expert's language, and the best way to find a suitable expression with the right degree of authority and unintelligibility is to use what we like to call a *buzz-word generator*. This kind of *nomenclature generation facility*, as an Expert would immediately call it instead, is a random yet infallible way of describing your proposal.

24

Number	COLUMN A:	COLUMN B:	COLUMN C:
0	synchronized	monitored	parameters
1	total	digital	facility
2	reciprocal	analogue	concept
3	systematized	management	flexibility
4	integrated	logic	programs
5	functional	correlative	option
6	incremental	balanced	hardware
7	parallel	optical	contingency
8	compatible	third-generation	mobility
9	viable	transitional	projection

THE NOMENCLATURE GENERATION FACILITY

Take any three-digit number, select the words from each of the three columns in the table, and out comes just the kind of term you need to transmute any ordinary-sounding proposal into an infallible and instantly-recognizable example of Expert language at its most refined. It is remarkable what a few *integrated third-generation programs (4, 8, 4)*, or a *reciprocal analogue option (2, 2, 5)* can do to an otherwise assimilable paragraph of prose.

Human societies have always liked to place faith in higher beings, and the series of gods, spirits and unknowable forces which have been erected by mankind over the millenia is enormously varied. But of course in recent decades the support for organized religion in Western societies has been distinctly on the wane, and Experts have stepped smartly into the vacancy.

There is now a clear tendency for actual worship of the Expert by the public. The money Experts are given, the power and respect they are accorded, and their licence to arbitrate over life, death and the processes of birth is a delightful recapitulation of the reverence that people used to reserve for deities.

Not all Scientists are Experts

A few decades back in history we saw the condescending patronage shown by the imperialist whites to the

25

coloured races; in the 1930s children's story-books embodied a deliciously well-studied attitude of superiority in the moneyed classes towards the lower orders, and even a decade ago it was possible to read short-sighted accounts of backward and unenterprising rich Arabs parking cars they hardly knew how to drive alongside their tents in the desert. How things have changed!

The cue for the emergence of Experts in the Western world came from the world of science which (if a little cranky round the edges) had always been solid and dependable.

It is perfectly true that scientists deal with terms which most people find foreign, but that is a reflection of two factors: *First*, the degeneration of education, so that words which any ordinarily-informed person might know are left in a limbo of ignorance for many. *Secondly*, the occasional foray by scientists into territory where specialist terms simply have to be employed.

However, there is no branch of science which cannot be explained to anyone, even without the convenient shorthand of the specialist terms (it may take longer, but that is all). And it is important that we understand why science falls back on 'long words'. It is to aid communication, to facilitate argument, and to provide a condensed method to transmit complex ideas in a short and assimilable way. All specialists use these terms. It is not necessary to look into the world of high-powered scientific philosophy, either; no-one who is not an experienced knitter could make head or tail of a knitting pattern without someone standing by to 'translate' until the shorthand had become quite clear.

Scientists and specialists are concerned with clarity and communication. Experts use long words to hinder communication and to repel all outsiders. It is no good hoping to find a helpful translator of the Expert's language, for there is no simple meaning behind those long words. There is, in the main, no meaning at all. The terms are there to impress.

26

A scientist uses experiments to substantiate a new thesis or to disprove an earlier one, but no Expert is there to solve problems. He is in the business of perpetuating them, for this is the only way to guarantee a well-paid lifetime's work. Arriving at new revelations is a very difficult task, but carrying out impressive-looking demonstrations is the easiest thing in the world if someone else has paid for a roomful of apparatus to do it in. So whereas the old-fashioned scientist uses experiments to create new ideas, the Expert uses complicated items of equipment to make his non-work look frighteningly unassailable. He simply selects whatever results he needs to support his latest pet theory, and that gives him all the status he needs. All he has to do then is feed the figures into a computer (which he has programmed himself to perform in exactly the manner he wants), collect the results, and proclaim that 'jogging causes halitosis' or 'too much sex produces intermittent deafness', to which the only sensible reply is 'Pardon?'

Once Experts realized what it was that made the public respect scientists, it was an easy matter to acquire this status for themselves. The liberal use of complex terminology, plenty of flashing lights and whirring machinery, a few bubbling retorts and condensers to set off the whole show, were more than sufficient to attract respect without any need to do the difficult, creative, part. Since the Expert sets up the systems he is going to use to test his pronouncements, failure is almost impossible. An Expert is always right, primarily because he says so.

The Expert Mind

What do you think is the great mysterious driving force that motivates all human behaviour? Perhaps you go for the power-lust theory, that insane craving to dominate and dictate to lesser mortals, which seems to haunt the corridors of big business, government and the broadcasting studio. Maybe it is sexual frustration impelling an individual to live out fantasies in a real world that does not approve. Possibly it

27

is an adolescent quest for identity. Maybe it starts in the birth trauma. Or perhaps it results from greed?

All the time you are casting around for an answer (and there are many large, expansive and well-funded institutes doing just that) you may find endless possibilities passing through your mind. So preoccupied do you become searching for an answer, that the most important fact of all is neatly avoided. It is that the question itself is entirely idiotic!

The people who make a living by posing essentially useless questions are Experts, and so great is our faith in them that we never stop to consider whether the question was worth asking in the first place. It isn't, of course. The single 'prime motivation' does not exist and if it did it would be quite beyond the capacity of our minds to identify it. Such notions form in the minds of narrow-minded Experts who are trained to think that way. It is as groundless as those costly artistic extravagazas enthusiastically supported by innocent public authorities, where people make displays out of rubbish, or construct little piles of bricks in meaningful situations (that word again).

Experts make their living by setting up non-existent solutions for an imaginary problem. The so-called 'instinct for survival', which has been so popular in wild-life programmes on TV and which explains in mystical terms how animals survive in the face of disasters like forest fires, is just one example. There is no such 'instinct' at all. An animal will not ordinarily choose to sit around being burned to a crisp when there is the alternative of running – fast – away from the heat. Put like this, it seems obvious. But dressed up in a wordy piece of unintelligible prose the same notion acquires authority and importance.

Running away from the forest fire becomes an *innate survivalistic potentiality* which is a derivative of the buzz-word mentality and could easily have come from a nomenclature generator of the kind we considered on page 25). The experienced Expert with an eye for gobbledeygook who noticed that the rabbits were running nose-to-tail from a burning wood might even demonstrate that it was *founded on an anal-fixated sado-masochistic fulfilment sublimation*, which ought to be worth ten years' research grants of anybody's money.

Every Expert knows that translating a term into Latin or Greek is as good as solving the problem altogether. Patients who go to the doctor with an unexplained fever experience enormous relief when told that it seems they are suffering from a classic case of idiopathic pyrexia. This says nothing more than the words 'unexplained fever' but it says it obscurely, and that is what counts.

Then there was the recent occasion when the press revealed that Experts had at last solved the mystery of the disease that was infecting salmon stocks. 'The illness, in which ulcers of dead skin form on the surface of the fish, has at last been identified as Ulcerative Dermal Necrosis,' proclaimed the news, an expression meaning (as far as it means anything at all) that ulcers of dead skin form on the surface of the fish. The mystique is every bit as powerful, and just as specious, as the all-French menu in a sophisticated restaurant which would sound so terse and ordinary in the English original (see next page).

It is the layman who looks at the leaning tower in Pisa and concludes that it will fall over one day, unless propped up. An Expert erects lots of apparatus, takes endless readings, and then announces with all seriousness that *the structural configuration has an inherent time-dependent instability factor implying a gravity-responsive structural collapse situation without a prosthetically lateralised stabilitationalistic facility.*

The Expert Analysed

The concept of a competitive, self-interested, incomprehensible Expert is by no means a new one. In 1726 Jonathan Swift's *Gulliver* described the Academy of Lagado, in which Gulliver found a raggedly-dressed researcher who had for eight years worked 'upon a project for extracting sunbeams out of cucumbers, which were to be put into vials hermetically sealed, and let out to warm the air in raw, inclement

29

Menu

Plain Language
packet vegetable soup

French Style
potage des légumes coupé fine assaisonne au manière technologique

Nonscience Description
Aqueous infusion of freeze-dried Daucus carota, Brassica rapa, B. rutabaga, Allium cepa and NaCl with mono-sodium glutamate

Plain Language
Bacon and eggs, chips and peas.

French Style
Lard grillé Americain et œufs du poele à frire avec pommes frites et petits-pois réfrigérés

Nonscience Description
Dicotylidean striated musculature, partially pyrolysed; heat-denatured Gallus domesticus ova; lipid-caramelised Solanum tuberosum with senescent cryogenic fruiting structures of Pisum sativum.

Plain Language
Bread

French Style
Pain coupé par tranches au maison

Nonscience Description
Farinaceous Saccharomyces cerevisiae growth-substrate

Plain Language
White coffee

French Style
Mocha au lait complet

Nonscience Description
Fermented Caffea arabica solutes with $C_{12}H_{22}O_{11}$ refined extract and sebaceous underbelly secretions of Bos bovis var; domesticus.

summers'. The man wanted funding and was keen to emphasize that it would take plenty of time before success might be attained ('eight years more' in this instance). Swift described how his creation, Gulliver, has difficulty understanding many of the people he meets in the academy, and the emphasis which the twentieth-century Expert places on endlessly publishing the same kind of material, over and over again, has an intriguing precedent in Lagado: 'I saw another at work to calcine ice into gunpowder, who likewise showed me a treatise he had written concerning the malleability of fire, which he intended to publish.'

One, more recent, description designates the Expert in terms of a 'paranoid, noisy, obsessively compulsive long-shot gambler who simply does not know when to leave well alone.'*

Elsewhere he has been categorized as 'being able to turn out, after innumerable punchings, an infinite series of incomprehensible answers, calculated with micrometric precision from vague assumptions, based on debatable figures taken from inconclusive documents, and carried out on instruments of problematical accuracy by persons of dubious reliability and questionable mentality for the avowed purpose of annoying and confounding a hopelessly defenceless operating organisation.'**

Paul Weiss, who put together the symptoms of this kind of Expert research as 'irrelevance, triviality, redundancy, lack of perspective and unbounded flair for proliferation', went on to describe 'bewildered youngsters composing research projects like abstract paintings: picking some colourful and fashionable words from the recent litererature and then reshuffling and recombing them into another conglomerate, yielding a stew of data both undigested and

*Kline, N.S., You Can't Get it from Here, *Indian Journal of Psychiatry*, 1: 18-25, 1959.
**Anthony, R.N., Budgets in Laboratories, *Management Control in Industrial Research Organisations*, Harvard University, 144, 1952.

indigestible. We see narrow specialists lavishing their pet techniques on reconfirming in yet another dozen ways what has already been superabundantly established to everybody's satisfaction.'*

Magnus Pyke, back in the days when he sounded more like a disorder of fish than a household name, lyrically wrote of: 'diligent modern researchers analysing 716 samples of bath brick and writing a report with 1002 references in the literature entitled *A Preliminary Communication on the Sesquioxides of Silicon as Determined by X-Ray Crystallography with Special Reference to the Jurassic Minerals of Central Basingstoke* without a glimmering of any of the general implications behind their immediate preoccupation.'**

Pyke describes this as the ultimate joke, but it is only that if contemplated from the outside. To an Expert there is nothing remotely funny about this complex system of public subservience, the cooperation of a bewildered government, and the need to acquire bigger and better arrays of equipment from which come kudos, acceptance and peer gratification.

Like all other power-seeking groups, Experts have no sense of humour. Bureaucrats the world over are a humourless race, which is why they have given us the choice examples of Expertistical writing which we, as spectators, find so delicately whimsical. Here is a passage that has all the hall-marks of a light-hearted spoof – the contorted syntax, the laughable verbosity, the unapproachableness; but no, it is a perfectly serious example from a leading scientific magazine.

Would the public interest be better served were a statutory obligation placed on both road authorities and manufacturers of road vehicles so that a claim to damage might lie where some accident could be shown to have occurred due to some feature of a road or vehicle which, while not as might be fairly be described as a defect in construction of the road or vehicle, was nevertheless a contributing cause in

*Weiss, P., Experience as Experiment in Biology, *Science, 136*: 468, 1962.
**Pyke, M., [in] *Nothing Like Science*, John Murray, London, 1957.

the ultimate damage or injury which reasonable foresight, applied against the background of the facts and technical knowledge might have avoided?*

There is no need to be long-winded to be unintentionally funny, however. What better example could you find than the child specialist who advised:

If your baby does not like fresh milk, boil it.

Then there was the paper on contraception which stated that 'nothing is less likely to appeal to a young woman than the opinion of an old man on the pill', and the classic rejoinder in a survey on suburban health which ran:

QUESTION: Do you smoke after intercourse?
ANSWER: I don't know, I've never looked.

Yet some of the most perceptive comments about the way Experts work have been written by people who *meant* to be funny, like Mark Twain when he wrote: 'There is something fascinating about science. One gets such wholesale returns of conjecture out of such a trifling investment of fact.'**

Sometimes it has been suggested that humour is so far divorced from the Expert's way of doing things, that it is almost an antidote for the effects of their own brand of Nonscience. 'Mankind is no longer able to keep track of his own achievements,' wrote László Feleki. 'Our life has beome so mechanized and electronified that one needs some elixir to make it bearable at all. And what is that elixir if not humour? It is decisive for the future of mankind whether humour and science can keep in step.'***

Naturally, it is when this form of light-heartedness comes face to face with the monolithic might of the Expert mentality that the fun starts. One of the more innovative scientific humorists was R. W. Wood, who once startled a guest by cleaning out some traces of

*Nature 195: 1244, 1962.
**Twain, M., Life on the Mississippi, 1874.
***Feleki, L., Impact of Science on Society, 19: 279, 1969.

dust in the optical tube of a telescope by pushing his cat right through it.

Wood, a distinguished physicist, created one of the greatest short-lived mysteries of biology. He lived for a time in a flat in Paris near the Sorbonne. Just one floor below, with a balcony immediately beneath his, lived a woman who kept a few pets. Among them was a newly-acquired tortoise of somewhat small size. She had spoken to Wood about it, and he advised a diet of endives as a means of accelerating its growth – it was 'one of the latest discoveries', he assured her.

On the same day, from the local market, Wood bought several tortoises of his own, ranging from a specimen about the same size as the new pet in the flat downstairs right up to a monster-tortoise all of two feet long. He rigged up a sling with which he could lower one of his tortoises down to the balcony below and exchange it for the present incumbent, whilst the woman was out during the day. He started to do this every day or two, until 'her' pet had grown at the most prodigious rate for a week and a half. By this time half the local Department of Zoology were following the developments with avid interest, and the staff were reportedly astonished at the rate of growth for the normally slow-growing species.

Once Wood had reached the largest tortoise in his collection, he confided in his bemused neighbour that a change of diet away from the endives might help the animal maintain more manageable proportions. She stopped the endives at once, and, by means of some secretive further exchanges, Wood worked his way back down the range until the original tortoise was safely back in its pen. Exactly how the ballooning animal resumed its original dimensions was never revealed to the astounded zoologists, and I am not certain quite what they made of it all in the end.

A similar prank was played by experimental physicist Carl Bosch in 1934, when he found he had an impressionable young science journalist living in the opposite apartment. Keeping watch with his binoculars, Bosch (acting in the guise of his own professor) telephoned the man and told him he had perfected a way of detecting visual images through a telephone line. The incredulous journalist obediently put out his tongue and pulled all manner of funny faces on command, all of which were faithfully reported to him back down the telephone line.

Matters only came to a head when the unsuspecting journalist sat one morning with his back to the window. When Bosch rang, he insisted there must be a fault on the line, and the gullible reporter was soon offering to help trace the short-circuit. 'We must see if the instrument is fully earthed,' snapped Bosch. 'Bring the fire-bucket towards you. Now hold the phone by its lead and gently, *very slowly I tell you,* lower it into the water.' Only when it was completely immersed and the line went dead did the truth begin to dawn on the victim. And meanwhile it is said that the local press were reporting the new discovery with more than a little excitement.

Perhaps the most successful illustration of the capacity the public have to believe that Experts are capable of almost anything was the hoax by Robert Locke of the *New York Sun* over a century ago. He found out that Sir John Herschel, the astronomer, had left for a prolonged voyage round the Cape of Good Hope where he was preparing to map the southern skies. Locke began to publish articles on the important 'discoveries' the expedition was making – every article being composed by Locke in the evenings after work, and none of them containing a shred of fact.

Sun readers were told of men, plants and animals that Sir John was observing on the moon, as he gradually began to reveal that there was a whole alien civilization up there. And did people take it all seriously? Indeed they did. The *New Yorker* emphasized that the reports opened up a new era in scientific discovery, whilst the *New York Times* published a leader which described the claims as 'plausible and possible'. A host of meetings took place to

35

herald the new breakthrough revelations in many parts of the world. The tale held water throughout the summer of 1835, and served to boost the circulation of the *New York Sun* to unprecedented levels before Locke was forced to admit he had fooled the world.

A number of more recent 'important discoveries' which have swept the world have been just as fanciful. At the turn of the century there were N-rays, which turned out never to have existed; and in recent years there was polywater, which had hundreds of eager camp-followers publishing interminable papers on the properties and uses of this newly-discovered substance which (with the best will in the world, for there was no question here of a hoax) never existed outside the fertile imagination of the research workers themselves.

How prophetic were the words of US Senator Simon Cameron back in 1901:

I am tired of this thing called science ... we have spent millions on that sort of thing over the last few years, and it is time it should be stopped.

It has been stopped. In its place is a new establishment founded on data and jargon, a breed of people dedicated to obscurantism and the acquisition of information for its own sake, cloaked with mystique and omnipotence.

This is what I call *Nonscience*, and studying it is vital if you wish to understand what is happening today, or to survive what might happen tomorrow.

CHAPTER TWO

Experts at Large

Did you know that eating over-ripe tomatoes increases your sex drive? That baldness makes you a better car driver? Or that the number of veins on the side of your tongue predicts the age at which you will die? Neither did I, and the best of my knowledge no such relationships have yet been put forward. They are invented theories, made up for the sake of this chapter.

But there are two points worth emphasising: *First*, it is surely only a matter of time before these views (or something like them) are put forward as perfectly serious; *Secondly,* they do have that ring of authoritarian accuracy and wisdom. They seem perfectly credible. Many aspects of our daily lives have already been altered by equally far-fetched theories.

This is all a far cry from what science is meant to be all about. Science is essentially a quest for truths, a drive towards understanding reality, a key to simplifying our understanding of the universe and our place in it. The new movement of *non*-science is quite different. That is a quest for power, and the drive is towards increasing obscurity.

Nonscience has developed out of science as germ warfare has grown from the field of microbiology.

But before we delve into its intricacies and the hidden delights it holds in store for its adherents, we should refresh our minds with a few examples. It was Nonscience which managed in Britain to get everyone out of bed, and dressed, in pitch blackness throughout the winter months by the simple expedient of abolishing Greenwich Mean Time. To be accurate, Greenwich Mean Time was not abolished completely, for it did remain in use for international timekeeping, space-probe navigation, the calculation of

37

astronomical constants and so forth. It was only for daily life in Britain that it was abolished (only in Greenwich, as you might say, was Greenwich Time outlawed).

That was one example from the public sector. But do not imagine that the world of science itself is immune from all this. In scientific research the most widely used unit of measurement for microscopic objects has always been the Greek letter μ, which can be spelled out as *mu*. But in a sudden flurry of reformist zeal, that has all changed. First the unit was renamed, so that it became known as a *micrometre*. Then it was redesignated by a new symbol, so that instead of μ it became μm.

Now, a micrometre can be confused with a micrometer, which is something completely different (a micrometre is the new unit of length, of course, whilst a micrometer is a gadget made of steel used to assess the dimensions of engineering components on the machine-shop floor or a slip of glass with finely-ruled lines used by microscopists). Then, μm looks rather like mu written backwards by mistake, and that causes endless confusion amongst inexperienced personnel in laboratories and publishing houses the world over.

Perhaps best of all is the fact that μ's most widely used abbreviation, the *micron*, has been dramatically banished. This put thousands of textbooks out of date at a stroke, invalidated innumerable measuring devices and items of apparatus, and condemned as inadmissible a term that was second nature to an entire planet-full of experienced biologists.

It is most important to realize that the change was entirely pointless. The old term was familiar and convenient; it was shorter, it was not liable to be confused with the same word already current in two other areas of contemporary practice. This is the hallmark of Nonscience at work. It creates work for the Experts by adding to the problems of others.

Take the revision of the wiring colours for electrical equipment in Britain which was phased in during the

38

1970s. In the old days, the three wires on an electrical lead were red (the live wire), black (neutral) and green (the earth). Now, these are such sensible colours and so easy to remember that they could have made a model for the world to adopt. Red for danger is almost world-wide, only the People's Republic of China preferring to designate red as a colour of positivism rather than as a warning. Black is a good colour for the neutral wire, too; no colour could be more 'neutral' than that. And green for the verdant earth itself is sensible.

Suddenly, as you might expect, the new colour for the live wire was that dull, neutral colour: brown. The neutral lead was standardized as blue, surely the epitome of a sparking live-wire. And the earth was redesignated as green and yellow stripes. A high proportion of the plugs that have been connected with the new colours are wired wrongly. Most people seem to remember that the earth lead is striped, though some people still think that a striped lead is livelier, and assume it should be joined to the live terminal. But the brown and the blue wires are always being confused, and when people do summon up some kind of mnemonic to tell the two apart, as often as not the blue lead turns out to be the live wire for very obvious reasons.

The Expert's answer to all this is simple. He proclaims that the change was all part of progress, and was designed to make it easier to export electrical goods. But that will not do. In the first place, detailed leaflets are supplied with all appliances fitted with the new colours, and that could just have easily been done with the colours remaining as they were. In that case the leaflets would only have been necessary for exported goods anyhow. Secondly, if it really was necessary to force everyone to utilize the same colours, then it would have been better to encourage the widespread use of the green/ red/ black standard, which is very much easier to remember than its arbitrary new rival. And thirdly, the exported proportion could just as easily have been wired with any other colours demanded at the time. International homogeneity is entirely unnecessary in a case like this. You could add to that the fact that voltages vary from state to state as it is, and if something as fundamental as that has to be taken into account then the matter of differing wiring standards could hardly matter less.

But such arbitrary, inconvenient and costly changes are what an Expert means by 'progress'. To you or me, this word has connotations of improvement, of betterment, of moving forward in some way. But not in the Expert world. There progress means changing anything that hasn't been altered round for a while. 'If it moves, shoot it,' used to be the motto of the great white hunter of the 'twenties. To today's Experts the wording has changed, but the result is almost as bloodthirsty: *'If it stands still, change it,'* they say.

Metrication and the Decimal Point

During the 1970s the British introduced decimal currency. Some people wanted the shilling (formerly of twelve pence) to be divided into ten pennies, which left both coins relatively unchanged and added a new unit based on the former ten-shilling note. The result of this three-tier system would have been that the principal unit of currency and its immediate subdivision would have retained the same value, whilst the smallest of the three (the penny) changed by just a few percentage points upwards. It would have been simple.

But that is no way for the mind of an Expert to work. Instead, the old division of the pound sterling into 240 pennies was replaced by a new division into 100 new pence (indelicately named 'pee'). In no time at all the economy was shattered. People who knew perfectly well that an icecream, say, always cost three pennies soon accepted three *new* pennies as the cost instead. Prices leaped up by the 240% you would anticipate. And this helped to undermine the whole global economy, which was still geared very closely to sterling.

40

It was a triumph of Nonscience at its most refined. No-one understood the new system, and the government Metrication Board quickly stepped in to squash any attempts people might make. I used to employ a simple 'doubling-up' calculation. Take a sum in new pence, double it, and express the answer in shillings and pence. Thus 25p, doubled, becomes 50, which we call 5s 0d. Worked out the other way, an old half-crown of 2s 6d becomes 13p (actually $12\frac{1}{2}$p, but near enough). In a rational world you would think that this kind of simple calculation would be pressed on everyone as a handy means of making conversions until the new system took over. But there is nothing in the least rational about Nonscience. People were told very strongly to attempt no such conversion. Just accept the new prices, it was said, and forget what they used to be. The result shook world trade, and is still doing so.

Stock Markets
FT Index 530.8 down 7.0
FT Gilts 64.06 down 0.67
FT All Share 311.98 down 3.16
Bargains 16,103

Sterling
$ 1.9475, down 75 points
Index 91.8 unchanged

Dollar
Index 105.7 up 0.8
DM 2.2212 up 75 pts

Gold
$402.50 down $6.50

Money
3 mth sterling 15¼-15
3 mth Euro $ 12⅞-12⅞
6 mth Euro $ 12⅞-13⅛

According to Napoleon, metrication in France ",violently broke up the customs and habits of the people, as might have been done by some Greek or Tartar tyrant". Protests were suppressed with prison sentences and bloodshed."

The casual observer who glances at international high finance would be forgiven for wondering whether 'decimal' notation really matters.

Decimal currency, of course, does not exist. In a

41

decimal system the pound would be divided into *ten*. But it is not. Instead, it is divided into a *hundred*. This is not decimalization at all, but centimalization. So the decimal currency revolution even got its own name wrong.

The introduction of metric units has been even sillier. To begin with, the decimal point is not used in any of the countries where decimalization is traditional. There they use the comma. To them, 1,250 (which to you and me signifies one thousand, two hundred and fifty) means one and a quarter instead. The confusion involved in reading printed data from books or from calculators is beyond measure. To take just one example, the laboratories and workshops of Rolls Royce in Derby get around half of their working drawings expressed in metric units and the rest in Imperial inches. A figure like one and a quarter milli-metres is always written 1,250 and said 'one comma two five zero', whereas the same figure as inches is said 'one *point* two five' instead. The metric system was going to be introduced as a means of cutting out confusion ... and here it is, working side by side with its Imperial counterpart, effectively doubling the work and quadrupling the chance of errors.

In money the main difficulty is that no-one is sure where that decimal point (centimal point, if you prefer to be accurate) should go. How would you read £12.34? That is easy: twelve pounds and thirty-four pence. But what would anyone make of 12.34p, for example? Twelve pounds and thirty-four pence as above, or twelve point three four pence?

In formal measurement there is said to be less chance of confusion, as the millimetre has been adopted as the base from which to work. This is an interesting unit, partly because hardly anything is ever measured in millimetres anyway, but in deci-metres or some other more manageable unit. The millimetre is unique as a unit of measurement in that many people cannot even see it. And of course when an every-day length is expressed in this form it becomes useless in two ways:

42

a) because the figures are inconceivably large, which makes them easy to misquote and impossible to imagine (thus the room in which I sit at this very moment is 3794.5 mm wide);
b) because the figures are also liable to imply a thoroughly spurious degree of accuracy which no-one in real life would ever claim.

The controversy between the comma and the point is not the end of the matter, either. Even if we confine ourselves to the way people write out sums of money, there are a good half-dozen (I do beg your pardon, 0,5 doz.) alternative ways of expressing yourself:

> £2.56 for typewritten copy
> £2·56 in printed matter
> £2-56 for some cheques
> £2 = 56 for other cheques
> £2^{56} for Americans to understand
> £2,56 in Europe

Finally, before we pass on from the vexed question of money, one added delight.

i) Before decimalization, centimalization or whatever you would like to call it, one English penny exactly equalled one U.S. cent.
ii) There were exactly 2.40 dollars to the pound sterling.
iii) 2.40 old pence were exactly equivalent to one new penny. What mathematical conclusion can be drawn from these interesting coincidences?*

As the years go by, greater chances for ambiguity are arising. We have seen, for instance, that the comma is employed by the Continental Europeans where a point is used in Britain and the United States (along with much of the English-speaking community of the globe). It is bad enough that 123,456 in French means an amount a thousand times smaller than the

* 'None whatever' is the correct answer.

43

same figures would mean in Britain. The trouble is that you can also find a point used in some countries to separate the thousands from the hundreds thus:

123.456

which in England would be equivalent to

123,456

and again that would be exactly a thousand times greater than

123.456

as we normally understand it. But this is where the problems come in. Suppose you wanted to write down numbers in sequence? You could always write, in words: *one hundred and twenty three, four hundred and fifty six*, etc. But if you did this in figures you would immediately be in trouble since

123,456

could mean exactly three totally different things. In writing large numbers in Western journals it is now usual to divide them into groups of three, thus

123 456

but that is just as you might list numbers in a separate sequence, as shown in the previous paragraph. In this way we have the following three conventions:

123 456

123,456

123.456

which separately mean at least *eight* completely different things. The variations if we permutate them are almost limitless. And it is worth noting that back in 1798 a Frenchman named Le Blond suggested using a semi-colon instead of a decimal point. So you could throw in this as a further alternative

123;456

and that should make it nearly impossible to make head or tail of the system!

In many ways this new era of hopeless confusion has had local implications, such as the swimming pool which was re-labelled:

3 metre (deep end)

.7 metre (shallow end)

and gave a nasty surprise to anyone with poor eyesight. Other examples were rather bigger, such as the metrication of the designs for houses on a Council estate which resulted in one entire building being squeezed out as the distances were 'rounded up' in the conversion process. Metricating the building industry has been possibly the most lunatic example of all, of course. In the first place the world's plumbing is based on Imperial threads and not metric ones. Secondly, you do not ordinarily export houses, so the benefits in that sense are hard to understand. In addition there is the fact that all the pre-existing houses in the United Kingdom were based on dimensions to the nearest foot, and the standard sizes of doors, windows, carpets, work-tops, sinks and everything else were changed so that they did not fit any of the buildings in which they were going to be installed. Finally, of course, the banishment of the standard brick size meant that new work on old houses didn't match.

At the time that metrication was launched it was stated to cost around as much as 400 new hospitals or over a million new homes.

The move towards centimal units looks even more absurd. Nature does not work on tenths at all, but, if anything, on the base of 2, and sometimes on geometrical or logarithmic principles. Cells divide in two, and so spores are often found in multiples like 48 or 96, for example. The figure 10 cannot be divided by as many numbers as the more rational 12; indeed there is no practical way of expressing 'one-sixth' or 'two-and-two-thirds' in decimal notation. The notion of a quarter, which is so easy to explain to a child, looks infinitely more intimidating as 0.250, and the banishing of fractions makes it hard to understand a whole range of important data. For instance, every film maker knows that, at 16 frames per second in a silent camera, each frame is going to be in the gate for 1/16th of a second. It is self-defining, and extremely

45

easy to calculate. The decimal equivalent of 0.0624 sec makes no sense to anyone, and throws a previously simple calculation into the realms of higher mathematics.

We have seen that decimalization is nothing of the sort, strictly speaking, but is actually centimalization instead. With that in mind it becomes doubly interesting to realize that fractions, by contrast, are perfectly amenable to changes up or down by one tenth. For example, 1/12th is exactly ten times bigger than 1/120 as anyone can see. Thus we can conclude that:

a) fractions are really decimal, whereas
b) decimals are not.

In any event it is difficult to use decimal notation in daily measurement. We count time on a base of sixty, so that 1.50 mins can either mean one minute and fifty seconds or one-and-a-half minutes, which is twenty seconds less. Worse still, some people use the base sixty for hours and minutes and *then* go to decimals instead of seconds. A time in the morning like 0830.45 is distinctly ambiguous.

But this is a dangerous topic, since some Expert is bound to come along and suggest the decimalization of the year. That is not an original idea, though; indeed our months of October, November and December (from the Latin *octo, novem* and *decem* meaning respectively eight, nine and ten) remind us of the Roman approach to the decimal naming of the months.

It is said that the first decimal-type date was noted down in an Indian inscription dating from 595 A.D. After the French revolution there was an attempt to introduce a Decimal Day with ten hours in the diurnal cycle, 100 minutes in each of the hours and 100 seconds per minute. They tried a decimal calendar too, with the year of the Revolution as one, but within a decade all that had been forgotten.

So we still have minutes and seconds, months and weeks. The foot is re-emerging in Denmark, the dozen holding its own in France and the Low Countries, the

46

pound (as the *livre* or the *libra*) holding on in France and Italy respectively. We cope with the Swedish mile and calendars, and with abandoned ease we deal with currency conversions as though they were as inevitable as sunshine or the rain. Is there any overwhelming need to reduce everything to a factor of 10? Yet the shadow of metric weights and measures is still hanging over the United States, and there are a few lessons that all Americans can learn from the British experience.

Meanwhile my guess is that the next Expert proposal is going to be for a division of the year into 100 equal parts. That means each new day, nd, will be the equivalent of 3.6525 old days or od. The increase is similar to the currency change which took place in Britain over a decade ago, and if the penny can undergo an increase in value of 2.40 times an inflation of the day by a factor of 3.6525 should be no problem.

The practical repercussions are varied. For instance, the new first day of the week (which we might call *Oneday*) would begin at midnight on o-Monday. *Twosday*, on the other hand, would not commence until o-Thursday after lunch. The new Decimal Year would give an increased working week, since 7 n-d equal 25.57 o-d, but a two week holiday period would be the equivalent of 51.135 o-d and that would be a very pleasant change. Even if you scaled down the vacation to one-half an n-month (by analogy with today's practice) then, since there would be ten n-month in an n-year, the holiday would still amount to 18.26 o-d, which (as all those Experts would quickly point out) is an increase of 30.4285714% on what we have today.

Out would go the haphazardness of counting in 60s (for seconds and minutes), 24s (hours), 7s, 28s, 30s or 31s (days), and old-fashioned dozens (months in a year) – and in would come a crisply regimented new system which would have everyone in a whirl.

Already in Japan there is a project to encode people by a 14-digit number, so that the poetic allusions to Miss Softness-of-the-Lotus-Petal and Mr

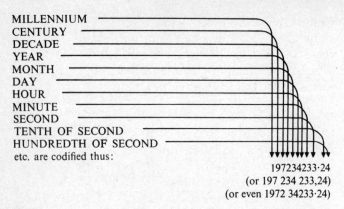

<pre>
MILLENNIUM ———————————————
CENTURY ———————————————
DECADE ———————————————
YEAR ———————————————
MONTH ———————————————
DAY ———————————————
HOUR ———————————————
MINUTE ———————————————
SECOND ———————————————
TENTH OF SECOND ———————————————
HUNDREDTH OF SECOND ———————————————
etc. are codified thus:
                                          ▼▼▼▼▼▼▼ ▼ ▼▼
                                          197234233·24
                                       (or 197 234 233,24)
                                       (or even 1972 34233·24)
</pre>

Data-system for a new decimal year: how to remember the date
and tell the time.

Might-of-the-Enraged-Tiger-at-Dusk are replaced
with data-codes instead: 29836478590218 or
98089875673894, for instance. And the Swedish authorities already identify everyone with just such a code,
so that any official can soon ferret out details about
your status, address and so on. With the new Decimal
Year, everyone could be labelled and slotted into a
data-system, and the new Expert approach would
enable the public to gabble like machines without
any need to resort to actual language at all. Since, as
we shall see in due course, there is every sign that
standards of basic literacy in the Western world are
dropping away as the cult of the Expert takes hold, it
is just as well that the future generations of video-
screen-watching automatons, which Experts envisage
we shall eventually become, will be able to communicate through data utterances, if nothing else. Then,
we will do away with such lyrical utterances as:

It must be about twenty-five to two on the afternoon of
January 5th, 1985 *anno domini* and I see my weight is ten
stone two; I stand (in my stockinged feet, as they say)
exactly five feet nine inches from head to toe.

and replace them with the new digital-era conversa-
tion which will sound like this:

48

Gone will be the need to put together phrases, or to think at all; and in a few complex-looking numbers there will be all the data anyone could wish for. A superabundance of figures, and not an idea in sight: just what every Expert dreams about.

What remains to change?
Clearly, it is only a matter of time before this kind of radical restructuring of our lives is foisted upon society by an up-and-coming Expert with a glint in his eye and little else. And no doubt we can anticipate the kind of progress that lies in store for the rest of us, through some intelligent understanding of the way Experts think. At the core of their unique mental processes lies an unremitting failure to identify cause and effect. If there is any single factor behind the success of Nonscience (I only said *'if'*, there isn't) then it must be this blindness to one of nature's most fundamental truths.

I can best illustrate it by two headlines:

CIGARETTE SMOKER SUCCUMBS TO LUNG DISEASE

TOMATO GROWER IN GANGRENE TRAGEDY

The first looks familiar enough, whilst the second does not. Yet if we study what might be the background to both cases it could well be that the facts are not as they seem.

CASE A concerns Willie Bombast. The lung disease to which he succumbed was actually tuberculosis, which he contracted from a consumptive lighterman with whom he had an unusually close relationship. The cigarettes were herbal cigarettes, prescribed by a naturopath in a vain attempt to cure his condition, and they contributed nothing whatever to his eventual fate.

The victim in CASE B was a steeplejack named Tom, who grew record-breaking tomatoes in his spare time. One autumn, whilst pulling up some weeds, he

49

overbalanced and fell against a dry tomato stem. From the splinter he developed gangrene, and that proved to have a fatal outcome.

So the sensible-looking association in (A) turns out not to be causally related at all, whilst the unlikely implication of (B) really *did* embody a cause-and-effect relationship: the tomatoes were the killers.

We have been lulled into an easy acceptance of causes even when they simply do not apply. If you find a young female victim of a coronary heart attack who has been taking the contraceptive pill, then it is so easy to assume that the one caused the other. But – in any particular instance – the causal relationship has to be proved (to put it more accurately, it is very hard to prove). But when you put two facts together like that, it is deceptively easy to assume that there truly was a causal effect.

Take: WOMAN ADDICTED TO TEA GIVES BIRTH TO DEFORMED BABY· No matter how open-minded you may set out to be, you can't help picking up the implication that the one caused the other. Experts make incalculable capital out of this kind of thing, and it is impossible to leaf through a scientific review without seeing a report in which some Expert is linking some behavioural manifestation, like car driving or sunbathing, with a disorder, such as piles or ankylosing spondylitis.

The entire process is thoroughly spurious. Suppose, for example, you looked at the age at which people started to drink alcohol in Britain and tied that in with the rate at which they grow. Your findings would look like this:

In a survey carried out on men and women aged from 5 to 50 years, it has been shown that those who do not drink any alcohol have growth rates up to fifty times faster than those who do. Statistics prove that once alcohol consumption becomes established, body functions alter until further growth in the community ceases altogether.

And there you have a clear-cut case for the abolition of all alcoholic drinks. The reason for the 'statistical correlation' is simply that people start drinking (legally) at around the same age as they happen to stop growing (18 years). Expressing himself in this manner, an Expert stands a good chance of starting a new branch of research with himself at the pinnacle.

The opportunities are limitless. In Wales there is a higher than normal incidence of spina bifida, for instance. The national vegetable in the Principality is the leek. So put the two together and you have:

The Welsh liking for leeks shows a close connection with the raised incidence of spinal deformities found in that part of the world, it was revealed today. Computer analysis of the latest available figures shows that the above-average leek consumption is closely related to the higher levels of spina bifida, Experts now claim.

And so on, until the leek is proscribed under some great anonymous regulation.

Some popular food ingredients have been banned because in large amounts they seemed to be associated with disease in animals. Cyclamates are an American example. They were fed in huge amounts to rats which showed a raised incidence of tumours. Critics argued that there were other items of the dietary intake of the animals which might have been involved, and that the dose rates were so high that it was difficult to know whether they could relate to the effects of the amounts people might ingest. The arguments went on and on, and are continuing still. What caught my attention was the fact that the banning of cyclamates led to a sudden rise in the amounts of white sugar and saccharine that were used to sweeten foodstuffs instead. Now, the evidence against white sugar has always struck me as somewhat flimsy, but that is not the point. It is widely accepted (and that is what matters in the game) that white sugar and saccharine have unwanted effects. So here was a piece of hotly-contested research, about one possibly hazardous item of diet being used to increase the public's intake of other compounds which were accepted as being, if anything, worse!

51

There have been many examples of inferred causality which proved to be baseless. For some illnesses (including pernicious anaemia) the 'septic focus' theory suggested that people with bad teeth were harbouring germs which caused their anaemic condition. So they had all their teeth pulled out, which added considerably to their problems. Tonsils and adenoids were routinely removed from children with simple sore throats (indeed many now middle-aged readers will be without their tonsils because of this operation). And in the 1930s the appendix was held to cause all manner of problems, so that prophylactic removal of the appendix before it had a chance to produce any symptoms became a new trend.

Shortly before the outbreak of the Second World War, in a London private clinic, a twenty-year-old woman student was taken in with abdominal tenderness, sickness and headache. She had the scar from an appendix operation across her abdomen, a neat little scar with a professionally-finished appearance. Drowsily she told the casualty officer that the scar certainly should be neat: it had cost her father some £90 for the operation, equivalent to at least £1000 ($2,300) in the early 1980s. Her condition worsened and eventually she was anaesthetized and the abdomen re-opened. There, much to the surpise of everyone, lay an inflamed appendix, distended with pus and a virulent source of toxins which had clearly threatened her survival. Within a few weeks there were two more cases of near-fatal peritonitis when inflamed appendices burst in patients who had long ago been operated on by the same surgeon. And then the truth came out: he had been making shallow incisions and then closing them neatly, without removing anything. He knew that the chances of appendicitis were statistically quite small (so that the likelihood of discovery was remote) and was smart enough to know that the fashion for appendicectomy was thoroughly pointless. So he found his own answer, by cashing in on the trend but leaving the organ in place.

Nothing was done, of course; in such circumstances

52

the infallibility of the Expert confers an automatic immunity to retribution even if it does get dented once in a while.

And where will the Experts move next?

There are many items in our diet which are known to cause human disease if vast quantities are ingested. They include: vitamin D, onions, kippers, wine, coffee and water. So those would be suitable candidates for an overall ban at some time in the future. Following the precedent of earlier experiments with animals, I could list a range of foods which cause death if injected intravenously into the mammalian body: bread, strawberry jam, liver pâté, fruit cake, porridge, butter and waffles are among them. Among the many foods on sale in the shops which have never been subjected to detailed examination for possible long-term side effects are: sandwiches, olde English tea-cakes, lemonade, stick-jaw toffee, sausages and french fried potatoes. Almost any novel food – from vacuum packed turkey roll and hot dogs to fish fingers and instant milk powder – could be banned in this kind of manner.

> **CAN CYCLING** ruin a man's sex life — or is it, on the contrary, an aid to potency? The question was prompted last week by a widely reported international difference of medical opinion.

Once in the swing of it all, you could find reasons to ban almost any aspect of our daily lives. People die from the exertion of making love, which can spread all manner of nasty diseases anyhow, so that could certainly be outlawed. The end result of pregnancy (a health hazard in itself) is leading to over-population and the trapping of bright and innocent women in households where they believe they are becoming drudges, even if they don't always feel like that, and this is causing international tensions. You could avoid the pregnancy risk by using the contraceptive pill, of course, but that produces enough risks as it is. There is a plain risk of cancer from drinking tea or coffee,

53

and the carcinogen 3:4 benzpyrene is present in any partly-burned food material including barbecued steaks, toast, bacon or browned potatoes. Coffee can produce pancreatic stress and that in turn is capable of exacerbating diabetes.

High level research into food safety always attracts public attention. Here the question is whether nuclear radiation changes flavour – an equivocal matter, if facial expressions are a guide.

At the meal table there is the chance of typhoid from eggs and carcinogens in kippers, while the artificial ingredients in bread are regularly condemned for a whole panoply of reasons. That popular seasoning agent, mono-sodium glutamate, causes funny fits in the susceptible; butter is associated with coronary heart attacks; tea can damage the veins; and salt (apart from being incriminated in all kinds of electrolytic disturbances of the bloodstream) is composed of sodium and chlorine ions: the former a dangerous metal that explodes on contact with water, the latter a legendary poison gas from the trenches of World War One.

At work, you are at risk on the factory floor from all those machines; if you are in middle management there are all the problems of competitiveness and restriction with the risk of duodenal ulcers; and at the top of the tree you are surrounded by heart-attack-inducing stresses. If you survive, there is the chance of a stroke.

What do you decide to do to avoid these hazards of daily life? Stay in bed? That will not do either, I fear. The incidence of ulcerating bed-sores is high. Even if you avoid those, hypostatic pneumonia and fluid stagnation in the lung are a real threat to anyone lying in bed, and plenty of broken hips have resulted from misguided attempts to get out of bed the wrong way.

Water itself is capable of threatening life, not just from drowning (far too obvious) but from cerebral oedema if you drink too much, and from the effects of chlorine on the intestinal wall if you don't. There is the fluoride question, too, of course, which will by all accounts condemn you to rotten teeth if it is absent, and all manner of nasty long-term effects if it is not.

Even the air we breathe could be condemned by an Expert at a loose end for something better to do. There is the smog, of course, and all those nasty pollutants; but a more adventurous approach would be to examine the gases of which air itself is made:

NITROGEN can induce narcosis which has been known to kill (and that makes up a little under four-fifths of air, or as the decimalization lobby would have it, 0.8);

OXYGEN leads to anaemia if breathed on its own, and can even cause pneumonia in experimental animals, so heaven only knows what it is doing to our children! (Oxygen is roughly one-fifth of the air we breathe.)

CARBON DIOXIDE (comprising 0.03 percent of the air) is a suffocating lethal vapour which extinguishes human life in a matter of minutes.

Interestingly enough, though these are the gases we all know about, little is ever said on the subject of

ARGON (nearly 1 percent of the air) even though there is over *thirty* times as much argon in the air as there is carbon dioxide! We know that there are rising levels of leukaemia, for instance, so that could easily be argued by an Expert to be the 'cumulative effect of genetic damage produced by this little-known gas'.

So for the Expert of tomorrow there is ample scope. With data from those computer memories, and with the willing acceptance of the public sector and the government agencies (if past performance is anything to go by), he could outlaw eating, drinking, falling in love, lying in bed and even breathing. It is a challenge which the student of Nonscience will find hugely entertaining. Doubtless all these new rules and regulations are just waiting around the corner, waiting for the Expert to strike.

CHAPTER THREE

The Fashionism Principle

Strange as it seems now, there was a time in the 1930s when the news media were full of reports that aluminium saucepans were a risk to health and should be banned as a matter of urgency. Earlier still, people used to worry about the smell of electricity that leaked from unused power-points and lamp sockets. In more recent years there have been temporary fads like these, and you may have wondered why a topic, which was so urgent for a time, suddenly ceased to attract interest and vanished with hardly a trace. Whatever happened to white sugar, which was threatening our survival? Where is that great debate about cholesterol, which filled the news media in the 1970s? What happened to the voiceprint?

> Voiceprints represent the human voice pictorially and are said to be as unique to each individual as finger-prints.

> sound spectrograms, unlike finger-prints, are subject to considerable variation; a person's voice changes with age and circumstance and can be disguised or made to resemble that of another person.

> There is no acceptable evidence to support manufacturers' claims that the voiceprints recorded by their equipment identify an individual almost as uniquely as do finger-prints.

As true to your individuality as fingerprints, voice-prints were going to end frauds and enable computers to recognize your voice over a telephone from a of millions ... but they too have vanished from the arena of public debate. Where is that great discussion over the pill, mouldy potatoes causing birth defects, blue asbestos which threatened us all in our homes, organic foodstuffs and 2,4,5T? And what became of the headline-catchers like space travel and the challenge of organ transplants?

These topics, and many more like them, seemed to be high-priority areas singled out for attention because of their importance, or because they threw new light on our predicament. It remains popular to believe that research projects are selected because of their merit, their timeliness or their significance to the future of mankind. In fact it is far less reputable.

Subjects for special attention rise and fall in the ebb and flow of fashion. A spectacular summit of interest, followed by eclipse – that is the fate of any topic when an Expert gets hold of it. The student of Nonscience knows this as *fashionism*. In choosing which subject to study or which aspect of society to alter next, the Expert must always be guided by the constraints of Fashionism, and his choice need never be tainted by considerations such as 'humanitarianism' or 'relevance'.

The term may remind you of the hovering hemlines of the dress designer, or the rise and fall of discs in the pop music charts. If so, this is a helpful coincidence, for there are parallels. All rely on the surge of carefully nurtured public opinion, disregarding practical considerations like quality or worthwhileness; all involve the redirection of large sums of money for the personal use of the star and his entourage.

You will find folk singers, who bewail the lot of the poor and under privileged, retiring to their lush penthouse suites after each concert appearance to count their takings from the over-priced tickets bought by open-eyed youngsters. The pop star poses in artificially-tattered jeans for the next album cover against a row of slums, whilst the air-conditioned limousine stands by at the end of the photo-call. And there are always plenty of Experts pronouncing against profligacy and pollution who jet from place to place in suits made from energy-consumptive artificial fibres and run the biggest gas-guzzling car for miles.

'Trends over the Decades'

1960s	1970s	1980s
Fallout	Cholesterol	Silicon chip
White Sugar	Vitamin C	Video
Cervical Cytology	(The) Ecology	Biotechnology
Fluoridation	Ozone Layer	Data Banks
Pollution	Technology Assessment	Wind and wave power
Conservation	Polywater	Immunotherapy
Pesticides	Recycling Waste	Digitalization
The Pill	Heart Attacks	Interferon
Marijuana	LSD	Lead Poisoning
Third World Countries	Underdeveloped Nations	Less Developed Countries

Do not be persuaded by Expert arguments that these trends are a reflection of real priorities, or that they reflect genuine merit. Fashionism does not act like that. When we knew little about the full extent of the risks from atom-bomb testing, there were worldwide demonstrations against fallout. In the 1970s the atmospheric tests started again, this time in a climate of opinion that was far more aware of the insidious threats from nuclear radiation. And did this give rise to correspondingly bigger and more urgent demonstrations? No – though our knowledge of the risks was greater, the charts had changed and fallout was no longer as fashionable a subject. Few people cared about it any more, no matter how much they should, and it seemed as pointless to demonstrate against the new generation of happy-go-lucky bomb testers as it would have been to scream for the Bay City Rollers or the Monkees. Now nuclear disarmament is suddenly back in fashion once more. I hope history will note the decades during which it remained beyond the pale, for (when fashions change once more) it will vanish as quickly as it came.

In the atomic energy field, the fashion in the late 1970s centred on power-stations, and huge effort was poured into proclaiming against the evils of nuclear power as an electrical generation energy-source. Everyone is now convinced that traditional sources of power would be better, and so support is being given to coal and oil-fired stations, many of which emit thousands of tons of sulphur dioxide every day of their operation, poisoning countless acres of verdant forest and killing countless people through chronic bronchitis every year. If nuclear power stations killed a single person there would be an immediate outcry; but nobody cares about the coughing thousands who are being poisoned by the traditional alternative. Such is the power of fashion. No wonder it matters so profoundly to every self-respecting Expert.

The heart transplantation boom is surely a classic example of Fashionism at work in its most rampant guise. It is surprising to realize that the first heart transplants were carried out in 1905. By 1961 the first implantable artificial heart had been built and tested, ready for human use. Then in 1963 a United States convict received a pardon in return for volunteering for a lung transplant. This operation requires far more extensive and radical surgery than cardiac transplantation, but even so it produced the merest flicker of interest and that was all.

The astonishing brouhaha that followed Dr Christiaan Barnard's experimental operation was quite unforeseen. Was the excitement due to novelty? Hardly, after the earlier examples of heroic surgery we have cited. Was it because the work was the culmination of a long series of research projects by an experienced team? No: other teams had carried out far more thorough investigations, using animals, and had obtained better results. Barnard has stated that the exact technique he was going to use only occurred to him on the morning of the operation, so it does not seem like a tried and tested procedure. Those who imagine that the operation on Mr Washkansky involved a new and strange item of apparatus would be wrong too, for Barnard describes the donor heart-lung machine as 'the old one of happy memories I had brought back from Minneapolis.' Even the operation, he admitted later, was not entirely free from 'stupid mistakes'.

The colossal international hysteria that followed the operation was a definitive example of fashionism at its best. Dr Barnard's meetings with film stars, statesmen, the Pope and countless other pillars of international society show how impressed people are by all this. There is nothing inherently less dramatic about such operations now (except that we know more about rejection, and can usually obtain reliable results in the right patients) yet the lack of public interest and the drop in the number of operations has been characteristic of those fickle charts and how a title falls as quickly as it rose in the first place. There has even been a backlash, as so often happens. The new fashion is *against* transplants because of the trendy debate about brain death. That too is an issue that has been current for decades and which attracted slight attention until it took off with a bang and rocketed to the top of the charts.

Plotting the rise and fall of topics in the Fashionism business is an amusing pastime, a veritable ongoing observational situation if ever I saw one, and those who took to heart the predictions of this book's predecessor will have been the only people who were not amazed when – right out of the blue – Metaphysics in all its forms from spoon-bending to telepathy suddenly peaked in the 1970s. The student of Nonscience could feel that it was on its way, but nobody else did. It is probably time that the scientific magazines began to print a chart like those you find in the pop music magazines, and the current position is summed up in my pioneering example.

There are welcome signs that one of the longest-lasting of the chart toppers, the women's movement, is on the way down. This movement was centred on the ludicrous notion that men and women were essentially identical. Fashionism has the power, not

61

Position Now	Last Chart	Title	Artiste(s)
1	8	Chips with Everything	Silly Con and the Modules
2	6	Culture Power	Mike Robe and the DNA Engineers
3	7	Have we got Sex All Wrong?	Frank Attitude
4	1	(You give me) Fever	Lassa and the Lab Escapers
5	10	What should we do with The Old Folk	Gerry Attrix
6	2	Here comes your coronary!	Hart O'Tack
7	5	Hard day's Graft	The Barnard Brigade
8	3	Close Encounters	U.F.O.
9	4	Birth pill Blues	Arthur Oma and the Clots
10	9	Beaches of Oil	Slick

WATCH OUT FOR !!!!

Position Now	Last Chart	Title	Artiste(s)
11	27	Ban the Bomb	The Fallout Brothers
12	8	Come fly with me supersonically	Concorde and the Detractors

STILL GOING STRONG !!!

Position Now	Last Chart	Title	Artiste(s)
13	10	Blue, blue asbestos	The Lung Shadows
14	6	Bent Key Boogie	Gellerites

The Research Planner's Top Ten Chart

only to take empty-headed topics to the peak of popularity, but also to influence the corporate criteria of a whole community until people begin to doubt the evidence of their own senses, and to lose touch with the most basic levels of reality. This particular example almost made it fashionable not to have babies anymore, and that is the ultimate example of the Expert's desire to triumph over the ordinary public. If it had worked, then the public – as a species – would have actually become extinct.

In some instances a topic has climbed steadily up the charts and has gained international acceptance, when it did not even exist in the first place. N-rays are a notable case in point. They were first discovered in 1903 by Professor R. Blondlot, head of the Department of Physics of the University of Nancy and a leading member of the French Academy of Science.

According to Blondlot, who already had a formidable reputation as an experimenter, N-rays were given off by many metals and in a darkened room they enabled you to make out objects that would otherwise be invisible.

A piece of paper with handwriting on it was a good example. Placed in a room that was progressively darkened, it could be watched continuously until the levels of illumination were just too low to enable the words to be made out. At the stage Blondlot would unleash his N-rays, and the writing could be seen once again. It was faint, but could be seen distinctly, he averred. His paper on the new discovery was published in the journal *Comptes Rendus* and it soon became a fashionable subject. The hypnotherapist A. Charpentier found that N-rays were definitely given off by muscular and nervous tissues, notably those of the brain, and his claims were authenticated by one of France's leading authorities on electricity and magnetism, D'Arsonval. By the end of 1903 a dozen articles and papers on N-rays had reportedly been published in French.

Early the following year, Blondlot announced the construction of a spectroscope which could analyse the wavelength of the N-rays. Another noted research worker, Jean Becquerel, discovered that they could be transmitted like electricity along a wire. By the summer of 1904 the total of publications on the subject had topped fifty. The French Academy responded by awarding Blondlot the Lalande Bequest of 20,000 Francs and the Gold Medal of the Academy for his discovery of this new form of energy. In September that year N-rays were a top talking-point at the meeting in Cambridge, England, of the British Association for the Advancement of Science. The subject was riding high.

But Professor Blondlot's magical N-rays did not exist. The truth emerged when our physicist friend, R. W. Wood, removed the prism which was claimed to be producing the radiation, whilst Blondlot went on claiming to observe their continued effects. So the saga of N-rays came to a traumatic end, and Blond-lot's claim to fame died a quiet and embarrassing death.

But this was an example from the turn of the century, when ideas were less critically scrutinized than they might be today and when, you may imagine, it was easier to make such a mistake in good conscience. If that is your answer to the episode then you are forgetting the saga of *polywater*. This example went further even than N-rays in terms of internation-al research and acceptance, and it dates from uncom-fortably recent times. The story began in 1961, when N. N. Fedyakin, a Soviet research worker at the Textile Technology Institue at Kostrama, observed small droplets of condensate inside fine tubes made of glass or quartz. He believed the liquid was ordinary water, and yet he reported that it was different from water in two important respects: its coefficient of expansion was higher than that of water, and its vapour pressure was lower. He then found that its boiling point was 150°C (compared with 100°C for normal distilled water) whereas if it was heated up to 700°C it thereafter reverted to normal.

By 1966 over a dozen papers on the so-called 'anomalous water' had been published in the Soviet Union, and it was then that the subject was discussed in Britain, at a meeting of the Faraday Society held at the University of Nottingham. J. D. Bernal, Eng-land's famous commentator and crystallographer,

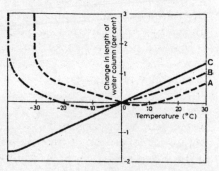

Figure 2 Unusual properties of polywater —curve A representing normal water, curve B a mixture of normal water and polywater, and curve C pure polywater, showing that it freezes into a glassy state at −40°C

began work on the material at Birkbeck College, London, and by the autumn of 1968 there were two British research teams and one in America investigating the new discovery. In 1968 Ellis Lippencott and his colleagues Stromberg, Grant and Cessac published some spectrum analyses which led them to coin the term *polywater* for the substance. They suggested that the conventional molecules of H_2O had rearranged themselves to form some kind of hexagonal lattice. Polywater emerged from this flurry of attention as having the appearance and consistency of 'Vaseline' and a density 40% greater than that of water. It was also found that at $-40°C$ it froze into a glassy solid that was 'not unlike ice' …

Within a decade after the first announcement of the new discovery there were over 400 individual research workers engaged on polywater investigations, and their conclusions began to get exciting. Many newspapers reported to an aghast public that it was now being claimed that polywater could spread its polymerized property to other bodies of water, turning them, too, into a viscous semi-solid. If a drop of the substance came into contact with the human body, for instance, then the water content of that unfortunate individual would turn to jelly and death would instantly follow. The property would then spread on throughout the rest of the world's water, causing a drastic and completely irreversible change in the properties of our planet's surface, and causing (a favourite Expert phrase, this) 'the end of all life as we know it'. At the same time the climatological and environmental catastrophe would be so massive as to make any hope of life re-emerging seem unreasonable.

It was not until 1973 that it became clear that the sticky, lethal liquid did not exist after all. A close search through the literature reveals that at the height of the craze, when hundreds of men and women of all degrees of eminence were experimenting with polywater, only some four or five were actually contributing to the expansion of the original hypothesis. Many of the others were simply jumping on the bandwagon, trying to collar as much funding as they could while the fun lasted, and resolutely endeavouring to make themselves into eminent polywaterists without bothering to pay attention to whether their favourite subject actually existed or not.

We have considered two categories of topic which became intensely fashionable, one of them real and the other imaginary. But there is a third, and that is the group of research findings which were not even a figment of someone's malleable imagination. The topics in this third group were actually invented.

> SIR — I very much regret to have to report that data published in the preceedings of the 28th International Congress of Physiological Sciences (Purves, M.J. 1981 Cerebral Blood Flow and Metabolism in the Sheep Fetus. *Advances in Physiological Sciences* 9: 199; 126. Pergamon, Oxford & Adademiai Kiado, Budapest) are false. I must also emphasize that none of my colleagues was involved in the preparation of this paper and the responsibility was mine alone.
>
> *Bristol, UK*　　　　M.J. PURVES

In the mid 1970s much excitement was generated by the claims of an American, Dr. Summerlin, who exhibited a number of mice in which he had undertaken skin grafts. The transplanted skin patches had come from black mice and were introduced into white ones. The transplants showed up plainly because of the colour difference, and even to the untutored eye the success of his experiments was plain to see. The news quickly spread that Summerlin had managed to overcome the normal immune response in the recipient mice, which in the ordinary way would have caused the alien skin specimens to slough off. Closer inspection, though, showed that the edges of the grafts were exceedingly smooth and regular, quite like the unaltered skin covering the rest of the mice. Little wonder: it then turned out that the black areas of 'grafted' skin had actually been inked in with a felt pen. The journals of science around the world, which had announced the success of the new method of obtaining successful grafting, were not amused. I was, though, and what is more I still am.

It concerns work by two American scientists, Mark Spector and Efraim Racker, that many researchers believed would unlock the mystery of why cells become cancerous. At least some of that work now stands revealed as a deliberate fraud, although it must be said that both Racker and Spector deny responsibility for the act. So excited were cancer researchers about the implications of this research that many of them still refuse to believe that all the work was dishonest. And indeed it is one of the ironies of many scientific frauds that it is impossible to say which results are faked and which are not, leaving a shambles that takes researchers much energy and time to sort out.

More recently the scientific journal *Nature* printed letters of complaint about the unreliability of some data collected in the physical sciences. The queries were passed on to the individual who first published them, and after a little probing he decided to make a clean breast of things, too; yes, he admitted, he had been so sure that his experiments were going to turn up the right answers, that he simply invented the data and dispensed with the need to obtain any real ones. The 'results' on which everyone had been basing their work were the figment of a fertile imagination and had never existed in fact. In 1980, Dr John Long of the pathology laboratories at the Medical School, Harvard, admitted falsifying conclusions in a research project on antibodies. The grant for the work, according to newspaper reports at the time, had been $150,000. Dr. Long resigned when the disclosure was made public in the *New York Times*.

It turns out that much of the work of the legendary Sir Cyril Burt on psychology and intellect was based on fiction, too. Burt's work was for decades regarded as a pinnacle of British research on educational psychology. He had spent much of the time since 1913 collecting data on twins and on the psychological development of the adult personality. With his two assistants, Miss Howard and Miss Conway, he accumulated files of data and in the 1950s and 1960s he churned out seminal research papers that were based on this immense fund of knowledge.

But all this changed in the fullness of time. Burt died in 1971, and over the following years commentators began to look increasingly closely at his legacy of

67

research. It did not seem to hang together as it should. Then in 1979 four separate categories of criticism of his work suddenly surfaced simultaneously and the journals began to take an interest. The categories of doubt were, to express it as politely as one may, well-founded and they did seem to cover almost every aspect of his life's work:

1 When interviewing families, Burt used to state that he had 'measured' people's intelligence, whereas he had actually guessed it by intuition and noted down his estimate as a hard and fast datum. This is a common ploy in making the data fit the preconception, of course.

2 Burt's figures were so reliable that he managed to produce the same results on different occasions from totally unrelated samples. Now, you might conclude that 'this just shows how good were his raw figures, and how accurate the models he used to manipulate them'. Unfortunately, the mathematics were too good to be real. The conclusions he drew were on occasions identical to three places of decimals, which is a statistical impossibility. It is as though you made three estimates of some stranger's age, one estimate based on the clothes he was wearing, another on the lines in his face, and the third on how he walked, and then concluded that each separate estimate suggested that the person had been born on 24 April 1935 at exactly 3 seconds after 4.15 p.m. It *might* be that your ability to guess someone's age and time of birth was as good as that, but I do not think that kind of result is going to impress anyone with any idea of how clever you are, even if it does do wonders for your reputation as a statistical fraud.

3 It became clear that Burt used to start with the results and work backwards. Now, that is certainly a deplorable thing to do; but it is so common in research work and data-gathering these days that it is hard to criticize anyone for that – least of all in the golden age of the cult of the Expert, when the whole tribe seem to spend much of the time doing just that.

4 Most telling of all, it was claimed that the two assistants (Misses Howard and Conway) were figments of Burt's imagination, invented to bolster his fictitious results.

Even Burt's supporters had a hard time salvaging

much out of that. So far as the figures were concerned, it really did seem that they owed most to a fertile imagination and an overwhelming desire to beat everyone else at their own game. Professor L. S. Hearnshaw, for example, has turned to Burt's defence in a noble attempt to shed some light of objectivity on the matter.* Yet the most useful service he could render to the 'grand old man' of British psychology was to explain where his peculiar characteristics, of bias, unsociability and a consuming determination to win, originated from. Hearnshaw finds two people who believed that the mysterious assistants might have existed, or at least that *one* of them did, yet his searches of the diaries of Burt from the late 1950s onwards (in which he noted every detail of his life, even when he went for a walk or could not find the cat) revealed no mention of them. Burt wrote to a Dr Nicholas of Evanston, Illinois in 1964 about 'the council's social workers, Miss Howard and Miss Conway', but they were not on the payroll of the London County Council at the time claimed nor were they listed on the staff registers.

Other inconsistencies appeared. In 1971 Burt wrote a letter (only a little while before he died) in which he explained:

… our own studies of separated twins were accumulated bit by bit between 1913 and 1939. We began with a very elaborate scheme including temperamental as well as intellectual assessments, and first-hand notes on home conditions. As my work increased I had to give up the time-consuming visits to homes, and the job of estimating non-cognitive traits. Most of the later cases (including a few post-war) were simply tested at school; results checked by the teachers … We did not bother with addresses.

which looks very balanced and objective. But some of his other letters belie this state of affairs. In 1955, he stated in a publication that in addition to the surveys carried out in the London schools, further data had

* Hearnshaw, L. S., Decline and Fall of Cyril Burt, *Sunday Times*, 8 July 1979

been 'collected by Miss Conway [who] thanks to numerous correspondents' had gathered much new material on twins reared in separate families. In direct contradiction of what he wrote in 1971, he went on (in a letter to Dr Lloyd Humphries at Illinois written in 1968): 'we have added considerably to the number of separated identical twins.'

In any event it is perfectly clear that Burt must have been cheating on the data somewhere along the line. If in 1955 he spoke of numerous correspondents providing further data, and in 1968 wrote of adding considerably to the number of cases, it is certainly hard to understand how in 1971 that has all become 'a few post-war cases', with the pre-war work proving too much to do as the 1930s progressed.

According to the survey of the Burt papers by Hearnshaw, Burt was often asked to provide details of his figures following a lecture or some other public airing of his findings. On almost every occasion, he did not provide them. Just twice he did: and instead of spending half-an-hour or so preparing a copy from his notes, his diaries apparently reveal that Burt spent two weeks 'calculating the data'. In fact most of his pre-war work was destroyed in 1941 during a German bombing raid on London, and yet Burt continued to 'quote' from those figures for years afterwards. Clearly he must have been re-inventing the results, for by then they no longer existed.

Sir Cyril Burt, then, was a great mind and a powerful figure. But he based his probing investigations on facts and figures which he invented himself.

There are similar cases where the hoax has been perpetrated by individuals who were out to trick the establishment, rather than impress. One example from the eighteenth century was the career of Johann Beringer, a Professor at the University of Wurzburg. He believed that fossils were not the remains of long-dead creatures (which was the view promulgated in the mid seventeenth century by Robert Hooke), but were specially carved stones made by the agencies of heaven. Beringer began to find them on his field

70

trips. Some had strange forms of life carved in the solid rock, whilst some others showed heavenly bodies, shooting stars and comets. Some seemed to have writing on them. One even said JEHOVAH. In 1726 he published a monumental work in Latin in which these remarkable finds were described. Unfortunately he then found a stone with his own name on it, and an enquiry revealed the truth: two of his colleagues had been carving the stones in their spare time. Beringer spent much of the next few years locating every copy that he could of his book, and burning them all. Because of the rarity value, the demand for the title greatly increased and an edition was reprinted 27 years after he had died. In more recent reprints, it is available to this day.

The twentieth century equivalent is the saga of the Piltdown skull, of course. A group of bones were faked up from artificially-aged fragments of human and chimpanzee skulls and were carefully smuggled into the finds from a quarry at Pilt Down in Sussex during 1911-1915. Among the quarternary gravels at the site were found fragments of skull, the right half of a lower jaw, and some teeth. Jubilant archaeologists named the 'new' species *Eoanthropus*, meaning 'dawn man', and hailed it as the earliest known member of the human race in Europe. Years later, when the truth came out, it seemed so obvious that the bones were a simple fake. Until the revelation, however, they were regarded (equally positively) as a genuinely important fossil. Such transitions in perception are frequently found in the Expert world.

Earlier this century a Viennese scientist was the centre of a controversy which remains unsolved to the present. He was Paul Kammerer, a student of the Lamarckian school of evolution which historically preceded Darwin's theory of 'survival of the fittest'. Kammerer set out to prove that the midwife toad, a species which normally mates out of water, could be induced to 'evolve' in the laboratory. The frogs and toads that mate in ponds develop special adaptations for the purpose; the males grow a characteristic

71

mating pad on the fore-limbs, with which to grip the female during copulation. Kammerer's view was that if the land-mating midwife toad would be induced to mate in water instead, it might develop mating pads over a few generations. Now, that is an improbable idea at face value. It was unlikely to work, and even if it did the results could be used to justify several alternative explanations.

But these objections to the theory did not occur to Kammerer as he set up an experiment in which the toads had to mate in water. Eventually he published his findings, and he revealed to a bemused audience that his male midwife toads had indeed developed black, horny pads on their fore-limbs. Unfortunately the experiments did not stand close scrutiny. It turned out that the toads had been injected with ink. The pads were a fake. At the age of 46, Paul Kammerer was a discredited researcher and he shot himself through the brain.

Not until 1971, when Koestler published a book on the subject, did the name of Kammerer re-acquire a little respectability. In Koestler's view he might not have been the fraud he was believed to be. Guilty or innocent, the celebrated story of Paul Kammerer and the midwife toad still shows that someone, some-where, was trying to poke fun at the research estab-lishment. Not for the first time, they succeeded. Not for the last time, either, I venture to add.

CHAPTER FOUR

Training for Today

There is an unyielding purpose behind the education and training that goes into making the Expert mind. Many people imagine that the purpose of training is to broaden the mind, develop a mature outlook, prepare for life, and that kind of thing. This is a fundamental error. The real aim is:

a) to provide work for educators;

b) to make students subservient to the establishment ('frightened of', that might almost be);

c) drain away any signs of originality, rebellion or free thinking in the process;

d) above all, to channel the mind into a path that takes it as far away from the needs of society as you could go without actually falling over the edge.

It is, as everyone will agree, a peerlessly successful aim which the modern school and college course, planned by an Expert, carries out to the letter.

Education or Training?
Strictly speaking we ought to avoid the word 'education' like the plague. Education is an old-fashioned word, like *love* or *the wireless,* which ought now to be relegated to comics and light entertainment programmes on TV. It is a word that dates you. The origins of 'educate' lie in Latin: *e-ducare,* to draw out. It has connotations of mind-expanding, experience-broadening, self-liberating mental development. It is now clear why the term has no place in the world today, and hardly any in a book about the cult of the Expert. There are few people these days who want anyone's mind drawn out at all: forced in, yes.

Shakespeare was one of the earliest to use the word in *Love's Labours Lost,* written in 1588. The word is a close relative, some philologists say, of *educere,* to lead forth. The contrast between this word and the simpler idea of teaching was brought out by writers such as Kingsley, who in 1849 wrote: 'The question is not what to teach, but how to Educate.' Properly educated, a mind is whole and free, independent and unassailable. Naturally, that is entirely out of place in the modern, thrusting world of the Expert and his obedient cohorts.

The death-knell of education was beginning to sound years ago. Wrote Mark Twain in 1900:

Soap and education are not as sudden as a massacre, but they are more deadly in the long run.

And more than two centuries before that, the great poet Dryden coined the immortal phrase:

By education, most have been misled...

which might almost be the by-word of the stalwart campaigners who, between them, have virtually extirpated any sign of that fuddy-duddy discipline from the high-powered world of today.

What goes on today is something entirely different, called *training*. The verb 'to train' derives from a Norman French verb *trahiner* – to drag (or to draw along). The derivation of a railway train gives an idea of this root of the word: in that sense it suggests being hitched up to the leader and dragged along those tracks, in distinction to a 'car' which can drive anywhere its driver wishes. That elegant and civilized language French used the word (as *trainée*) in contexts such as the dragging of carrion along a trail to leave a scent that would attract wild beasts for a hunt; there is no equivalent use of the word to mean 'teaching students' at all. In French I guess the word would imply dragging them along at the end of a rope so that they left a trail that others might follow... which, put that way, is close enough to what goes on in the class-room.

74

In Schools

The most important point to remember about school training is not so much what it does teach, as what it doesn't. Nobody is ever taught much about what they need to be successfully independent and contented young adults. To attain that end, and also to enjoy school you would need to teach children a range of things, including:

+ Trends in pop music; economics of the record industry, gigs and concerts; playing instruments; social themes in lyrics; development of pop as art-form, etc.

+ Choosing a house; mortgages, rent, rates and heating costs. Organizing alternative systems from communes to nuclear families; the law on occupancy; squatting: as squatter/as householder; decorating your home; management of a domestic economy.

+ Newspapers. No day should begin without a discussion of the news; a look at weather-maps; best-seller lists; political trends; cartoons, their style and content; and policies in a changing business world.

+ Travelling cheaply, and well. How to obtain air tickets that are far cheaper than you'd otherwise buy; concessionary rail-fares; hostels; hotel etiquette, dress and security. Cultures round the world; hospitality; how to be a guest. Helping to smooth bureaucracy *en route*.

+ Personal relationships, love, infatuation, friendships that come and go. Group relationships, peer-groups, emotional crises; hormonal problems of adolescence, what those overpowering feelings of emotion really are (this replaces the modern poke-and-jerk mentality of the Expert mind, which is utterly irrelevant to frustrated and frightened teenagers).

+ Areas of ignorance: what we do not know about the world. This includes almost everything of importance, for we do not know anything significant about the great 'killer' diseases, from strokes, heart attacks and obesity to cancer and arthritis (which does not usually kill people, but does the next best thing). What we do not know about food, plant and animal life, growth, death, predicting the future, economics, human destiny. How to be motivated to develop a probing mind in the area of research, pressing back those

75

boundaries of human knowledge that we used to read about in the good old days, when the fact that there were such boundaries was still admitted.

+ Drug addiction and habituation, from marihuana and television to tea, cocoa, coffee, alcohol, heroin and cocaine. What drugs do, what they are, how society has become founded on addiction to acceptable things (like tobacco) which are often far more dangerous than outlawed substances (like hash). How addiction to television stops people asking too many questions, and hypnotizes them into abject submission during the hours they're not safely in work.

+ After work: enjoying life, experiencing your existence to the full, raising goats, chickens or the roof on a Saturday night. The theatre, stage and outdoor life. Self-sufficient gardening, water sports from boating and swimming to deep-sea exploration. Hang-gliding and parachuting for beginners. Reading (including all the old favourites of the young from *Just William* and *Biggles* to comics of the *Superman* type). Explanations of how popular literature reflects the culture in which it was written. A bunch of academics now call this 'structuralism' I believe, but those of us who nurture an interest in cultural pressures have taught the principle for a great many years.

+ Family life. Liberation in parenthood. Exploring with children, watching and guiding their development. The history of an amazing era when people who were free from the drudgery of daily office work and who could go out and about with their families whenever they pleased, were called 'cabbages stuck at home with kids', and how much fun it is.

+ Making things yourself, from automobiles to aircraft, and including a whole course on current fashions in dress. For punks, safe ways of dying hair; for head-bangers, how to rivet patterns into leather; for motorcycle freaks, how to produce your own accessories. Weaving, home brewing, producing independent publications, book-binding, etc.

But hold hard there, we are becoming carried away by all this frivolity. That is an idealized school I write of, the kind of place where I would not mind sending my children and, what is more to the point, where they would not mind going, either. Such a school

76

would be a centre for the community; it would be an honour to teach there, and any time a famous person came to the area he would naturally drop by so that everyone could meet him for a spell. There would be children and old folk discussing traditions of the past, musicians practising pop music, printing presses that the kids had made themselves whirring away; and (as the Expert fraternity would be quick to have me emphasise) it would be a dreadful prospect. A free school – perish the thought.

You will feel safer if you return to the more familiar and ordered irrelevance of Nonscience. For there is a fundamental flaw in the free school model I was speculating about just now: it is relevant to today. Experts know that no training scheme should be relevant to anything except the protection of the status of Experts everywhere.

Nobody is meant to like school. On the contrary, it is intended to make you used to going to a place you hate, so that you can stomach the world that lies in store for you. The last thing you should do in a school is have fun. There are the mathematical Experts, whose *forte* is dry and academic figure-games and whose purpose seems to be the divorcement of maths from everyday life in as many cases as possible, and the obscuring of what remains; while other Experts conspire to make literature tedious (which will in due course discourage people from finding much out for themselves) and are always happy to experiment with 'new' ways of teaching languages, so that the child is increasingly bewildered by it all.

Experts have two goals in view:

FIRST, reducing the standard which the individual attains.

This is easily documented. By the 1860s, 90% of the British population could read and write, and the percentage was increasing steadily. There was a high level of political and cultural interest amongst the young population. By the 1970s, only 85% were literate, and the percentage shows every sign of decreasing. At that time (in the mid 1970s) the

77

number of pupils per class had reached 24 (in the primary schools) and 17 (in the secondary schools) compared with over 100 which was quite common a century before. This, of course, means that there had been an increase of 500% in the jobs available for teachers, a cut in work-load to a mere 20% of what it was – and a lowering of standards at the end of the day.

SECONDLY, giving the impression that the changes Experts have made have resulted in amazing progress.

In fact, levels of attainment are being progessively lowered. This is happening at two levels:

i) Examination passes are being awarded to less able pupils, both by lowering the pass mark and by adopting a more limited syllabus combined with a more generous scale.

ii) Selection procedures are being made easier, to meet the falling marking standards half-way. Thus courses that used to require several good exam passes will now do with two, and the grades they will accept are often Ds instead of the Bs and Cs of a few years back.

These days, for the first time ever, people get into college with less to show for their studies than was the case in an earlier age. The old-fashioned idea of improvement with time has been abandoned. It is part of the Expert notion of 'progress' once again: you must be different, that is all, not better.

There are students qualifying as teachers who never passed a single higher school examination in their lives, and it has even been reported that the bulk of mathematics teachers in junior schools have not passed maths. Perhaps the time is coming when all you will have to do is sign a declaration of subservience to the cult of the Expert. Sceptics might be forgiven for thinking that this is what we see happening today.

Could it be that the all-time-low standards of literacy and numeracy are an achievement? A show of strength of what Experts can attain and a step towards the state when people are unable to recognize that matters are getting out of hand? Every Expert, in his heart, dreads the prospect of an informed and critical public challenging his aims. In an illiterate and non-numerate society his position is that much safer.

The figures on reading and writing are, as we have seen, a clear indication of the falling standards. One could argue against them by suggesting that the techniques used to assess literacy had changed over the years. I have not yet come across the Expert who says so, but I fully expect someone to produce a survey filled with data that prove conclusively how literacy standards have actually improved. Mathematical arguments are easier to follow, however, and a survey published by the Institute of Mathematics in London* is one clear example. Their sample was 8,500 schoolchildren in the 15-16 year age-group. One of the tests involved 12 simple questions, such as *divide 24 by 6* and *what is 7 times 8*. How many of the pupils in the average Inner London class could get them all right? Only one. There were 1,400 pupils in the sample and out of the total, only 18 boys and 2 girls could get full marks in a test of basic, everyday maths with questions like, *If milk is 12p a pint, and I use 2 pints a day, what is the weekly milk bill?*

One recent initiative in Britain was the 1967 change in law which meant that pupils had to stay on in school for an extra year. That prevented them from maturing outside the closetted environment and also gave an extra year for those standards to be lowered. There are occasional complaints about the uneducated school-leavers we see around us. A report by the Confederation of British Industry published in 1976 showed that employers were becoming concerned at 'the many school leavers who have not acquired a minimum acceptable standard in the fundamental skills involved in reading, writing, arithmetic and communication'.

* Report by Institute of Mathematics, London, March 1978.

Two years after that, the Department of Employment's *Gazette** pointed out that people preferred to take onto their staff married women looking for a job, rather than school leavers. Presumably they represented an era where standards were higher, whilst the contemporary school leavers were unable even to communicate properly.

As matters stand, it will not be long before pupils are subjects of the electronic era. They will be able to stare into that flickering glass screen and take in simple instructions. Naturally they will be able to press buttons on calculators and read off the data that emerge. With luck, they should be able to follow instructions based on the time indicated by the watery digits on the dial of a watch. But (and this is where the Expert's might brings him to the fore) they will be in no position actually to communicate with each other, they will not have any intellectual equipment to challenge the Establishment or even to recognize when society is degenerating into nothing better than a feed-stock for the Expert's experiments, and their desire to change or to better themselves will have been effectively eliminated.

Rhodes Boyson, who was the Conservatives' spokesman on education when they were in opposition (but was not made Minister of Education when they were back in power, which may or may not be significant) wrote in 1979:**

The 12-year old school leaver of 1869 was generally literate and numerate … In 1978 it would appear that a considerable section of 16-year-olds have inadequate literacy and numeracy standards, many are not keen to work and they are trained not for apprenticeships but for a career of truant idleness.

Dr Boyson uses the work 'curricilae' which (since

* *The Gazette,* Department of Employment, London. January 1978.
** Boyson, R. Compulsory state education raises education standards? [in] Duncan, R., & Weston-Smith, M. *Lying Truths,* p.68, Pergamon Press, Oxford, 1979.

the plural of the word *curriculum* is *curricula*) suggests that the process is working even on him.

If school training brings out the capacity to abandon 'thinking', it is also destroying the far more basic ability to 'learn'. The alphabet and multiplication tables are no longer learned by heart. Biology students do not learn how to use microscopes, probably because that is considered a wee bit too adventurous for the modern mind.

I came across a third-year University student of biology who had never looked down a microscope, which I imagined set some kind of record, but which actually turns out to be entirely normal.

The problems of adaptation are many, no-one could deny that. But for the obedient mind that mutely follows the lead set by the trainer, there is always that goal of the printed tokens of obedience – paper qualifications, they are called.

Much has been written on the plight of the tyro, faced with this adapting, and here is one eminently quotable example:

'The newcomer as a stimulus or press in Murray terms is more likely to be perceived by open groups as compared with closed groups as a potential group resource with reference to group productivity rather than a restrictive and/or disruptive supernumary with reference to existing patterns of social cathexis and interdependent task functions. *Or perhaps more simply,* the newcomer enters a far more complex, interrelated system of elements in the closed groups than in the open groups, and assimilation is accordingly accomplished with reduced facility and increasing hostility, at least during the initial phase of the assimilation process.'*

The italics are mine, but I do hope that's clear now.

Experts know that it is through their calling and their rigorous training that fulfilment and above all, power, are found. For ordinary mortals, as Eder has

* Ziller, R.C., Behringer, R.D., & Goodchilds, J.D. *The Minority Newcomer in Open and Closed Groups.* Office of Naval Research, U.S., Contract ‡2285(02).

81

argued, the pattern of life is altogether different:

We are born mad, acquire morality and become stupid and unhappy. Then we die*

Schools for Scandal

There is still ample room for change in schools. It should not be imagined that the quasi-liberal lunacies of recent years are the worst excesses of the cult of the Expert. And it is important to understand the extent to which schools – apart from providing the raw material for development into tomorrow's Expert brigade – also offer limitless scope for the machinations of Nonscience at its most extreme. This explains why it is that schools come in for so much Expert attention: there is an essential feeling of patronage between the breeding-ground for tomorrow's Experts and the same establishment as an 'experimental animal' for today's Experts to play with.

Most individuals accept that if you spell 'spell' *spel*, you are an illiterate oaf. But if it comes to 'committee', 'embarrassing' or 'exaggerate' people are just as inclined to write 'comittee', 'embarassing' or 'exagerrate', and then shrug it off, and casually suggest that *you don't need good speling to comunicate efectively*. One classic example of this was a large advertisement for representatives which was published in the British quality newspapers, and was headlined

IF U KAN SEL
U DONT NEED TO SPEL

We believe selling is the art of verbal communication and therefore, back people not paper, the results not the spelling. We look for trainability, natural intelligence, high activity and the keen desire to succeed.

* Eder, P., *British Journal of Psychology*, 35: 81 (1962)

All I can say is that the time is around the corner when we will say to all potential Experts that: 'If you can con the public you don't even need to add up'.

REMEMBER
Spelling is not necessary to be understood
Calculators take the drudgery out of mathematics
Who needs alphabets, with computers to sort out data?

Many young school-leavers today are unable to work out everyday sums, to find telephone numbers which are alphabetically listed, to write coherent brief letters, even to say what they mean. Yet this merely makes Experts seem more important. And if a child cannot work out the simplest sum without a calculator, it does not matter either. You or I might say: 'What are you going to do when your calculator is at home or when the batteries run down?' whereas Experts know that their advice gives a fuller sense of security: 'If you get stuck, ask an Expert' is *their* answer.

In many ways you can justify some of the recent changes in teaching methods. For example, the theory of sets, which has bedevilled parents since the New Maths came in, or the *'Look and Say'* reading classes, have plenty of reasoning behind them. What tends to happen, though, is that all the old methods are swept out and all the new ones rushed in before anyone (least of all the bemused teachers who are often the victims caught in the centre) has time to get used to what is happening.

Take reading as an example. In the old days, there was much emphasis on phonetic pronunciations so that C-A-T spelt *cat* and D-O-G spelt *dog*. That approach breaks down rather when you consider that W-O-M-E-N spells *wimmin* or T-O-U-G-H spells *tuff*. English, although it is grammatically the easiest language in the world, which is why it is the international language in every field of endeavour, is utterly ludicrous in its spelling and pronunciation as words like TOUGH, THOUGH, THROUGH, COUGH, CHOUGH, THOUGHT and SLOUGH amply demonstrate.

83

Look and Say doubtless reflects the way adults read, of course. It does not apply to children, who have enough* problems telling the difference between a C and a G, or an I and an l. For them the phonetic method is obviously best at the start. Experts do not care what is merely 'best'. For them all that matters is what is new, or what happens to be the fashion. So when *Look and Say* came in, everything else went out. The child looks at the whole word CAT, sees the picture above, and then is supposed to memorize all those strange lines in one fell swoop, poor little chap. You could easily substitute some other wording (such as m-a-n-g-y t-o-m or whatever else took the fancy) and have the child read that out as 'cat' instead.

Do trendy revolutions in study methods upset the teachers? Of course. Does it make reading more difficult for the average child? Certainly. Would it not be better to use some *Look and Say* method at the right age to complement traditional phonetic teaching? Doubtless. Is it not confusing to experiment with abortive innovations like the *Initial Teaching Alphabet?* Definitely ... but there is no point in being so short-sighted. Children grow up far better fitted for a life under Expert rule when confusion sets in early, and for the world of tomorrow that is important.

At the heart of the comprehensive ethic, which has spread from the United States to Britain and beyond, is the creed of egalitarianism. The people who preach this dogma do not believe it for an instant. When they want an airline pilot, a surgeon or a teacher for their children, they would not dream of picking someone by democratic vote; they want the best and most skilful person available. When it comes to illness they head for the best specialists money can buy. But that does not stop them forcing comprehensiveness on their hapless fellows.

Before long you may meet an Expert in a cocktail party situation and ask him how he became so successful, only to hear him reply:

* NOTE: Another word to go in that list.

Apart from my essential humility there are three reasons for my peerless success. First, my unassailable grasp of maths. Third, my unbeatable memory. Unfortunately I forget the other.

And he will mean every word of it, quite sincerely.

CHAPTER FIVE

Success At All Costs

In recent years there has been much talk of a 'swing from science'. According to one official report edited by Professor F. S. Dainton*, there are some clearly recognizable reasons for this including:

+ rigour and unattractiveness of the school curriculum,
+ heavy factual content of training,
+ unimaginative presentation,
+ the fact that a body of received knowledge has to be acquired before speculation and imagination can be given free rein,
+ long hours of factual accumulation,
+ lack of contact with human affairs,
+ insensitivity and indifference,
+ the posing of more problems than solutions,
+ narrow concentration of studies in a given field,

all of which engender what the report calls 'repugnance'.

To the student of the cult of the Expert, these apparently contentious points are exceedingly elementary to explain. Experts *rely* on these very facts. To them such 'criticisms' are inexplicable, since they are what the whole edifice of Nonscience is built upon. The so-called 'rigour' of the curriculum and its 'unattractiveness' are what Experts have striven for over a prolonged period.

As university students will tell you, the basis of their training is designed to keep them as far from real people as possible. Many universities are built in

* Dainton, F. S., *Enquiry into the Flow of Candidates in Science and Technology into Higher Education,* Council for Scientific Policy, London, Cmnd 3541 (1968).

concentration camps well out of reach of the town. Others foster a heady sense of snobbishness so that the students feel it would be beneath their dignity to bother to have contact with the locals.

There is no real 'insensitivity', Experts argue, just a naturally superior and all-wise aloofness; no 'indifference', only a quality of studied omnipotence. And as for the criticism that Experts may be posing new ethical and moral difficulties – that is what being an Expert is really all about.

Yet still people muddle the two different aspects. I see the Dainton report referred to the fact in these words:

It is most important not to equate intellectual rigour with excessive reliance upon long periods of routine experiment, upon reiterated formal exercises based on elementary theory, and upon the committing to memory of large quantities of factual information which can readily be derived from basic principles.

Indeed it is *vitally important* not to confuse the two: what Dainton calls 'intellectual rigour' is nothing more than a polite term for *science*. Nobody could possibly confuse that with 'long periods of routine … reiterated elementary theory … memory of basic principles'. These phrases belong to Nonscience. No-one should confuse that, of all things, with science!

Let me illustrate the difference.

Intellect
This is the faculty which used to be called 'reasoning' or the ability to 'think'. The mainstay of today's training in Nonscience is not intellect any more, but memory. The aspiring Expert is given packages of programmed fact-sequences to learn by heart. He is instructed in a predetermined and precise fashion and the outcome (instead of being unpredictable and independent) is absolutely routine, utterly reliable. The use of memory as the yard-stick is also one of the central themes of the examination system, which makes testing potential Experts even easier.

Originality

Nonscience prospers because of its adherence to codes of fashionable facts. The idea of originality is that one should cultivate the heterodox notion whilst discarding the strictures that held earlier workers back. No Expert wants to be labelled as heterodox, his ways must be painstakingly copied from his forebears or the whole system begins to show instability. And as for the label 'eccentric', so often employed to describe scientists, every Expert will quickly point out that the term means (literally) 'being away from centre'. The successful Expert could want nothing less: he has to be exactly *in* the centre, where the decisions are made. He would never agree to any description that suggested he was a hanger-on, a fringe man, someone away from the action.

Creativity

As an abstract virtue, creativity has no place in Nonscience. Merely to create something is narcissistic, Experts say, when there are so many already-existing aspects of society just itching to be banned, changed or restructured. Instead of aimless forays into fantasy, Experts encourage increased concentration on working from set premises, following routine procedures to the letter, obeying explicit instructions and generally helping along the Nonscience movement.

Used in the form: 'Leave him alone or he'll only create' means 'do not tease him, or he will only *make trouble for you*'. With such a clear indication of the unacceptable face of creativity, who can blame Experts when they insist it is discarded?

Integrity

The mediaeval implications of the whole idea of all-out honesty clash violently with the Expert's need for caution, circumspection and guarded self-interest. If people were to start talking about drawbacks to the schemes Experts introduce, or confessing easily when they had made a disastrous error, then faith in Nonscience would begin to dwindle. No Expert ever likes to admit when he is wrong, and he covers his tracks by insisting instead that someone else has caused the problems, or that the idea which is proving so undesirable was not actually his in the first place. For today's high-speed world, circumspect ingenuousness is a far better approach than trying to reveal difficulties that would only worry the public, and might damage the research. Furthermore, some Expert projects have been founded on pure invention (and sometimes malicious selfishness), and where would anything as out-dated as honesty come into that? The new approach can be assimilated at a glance from the table:

TRADITIONAL TRAIT	CONTEMPORARY QUALITY
Intellect	Parrot-like memory
Originality	Strict conformity at all times to Lofty Principles
Creativity	Perpetrating the aims of Experts everywhere
Integrity	Cautious circumspection

Traditional-vs-Modern qualities in education
By this time you would be justified in wondering whether it was worth all the effort involved in becoming an Expert. There can be no doubt about this. People have enormous faith in an Expert, once appointed, no matter how un-Expert the person might otherwise seem. A mindless windbag whose work is mistrusted by everyone from the chief technician to the cleaner can become a peerless authority if he is asked to review a project for some hapless youth. An unqualified part-time reader can throw out a

manuscript (no matter how good) once he has been asked for his opinion by the publisher.

In domestic circles people appoint Experts at the drop of a hat. 'Why don't you ask Fred?' they say, 'He's the Expert', whether it is a blown lamp or a row of drooping cabbage plants that need replacing. If you were to go into a bar and mutter that you knew, for a fact, that tall blokes had longer willies than short ones, you'd be rightly scorned. But the same observation couched in the right way becomes perfectly acceptable.

R. Fischer: Penis Length and Body Weight.,
Proc. Biol. & Med. Sciences, 67: 103.(1964).
Re-evaluation of reliably measured data (N = 49) indicates a positive correlation between human penis length and body height. Acknowledgement: we are also indebted to Dr S. Dinitz for his kind assistance.

In the business and planning world Experts and their anonymity are utilized to take decisions that would otherwise be personally embarrassing. You pass the choice over and then use the fact that the decision has been taken on Expert advice to justify whatever happens afterwards.

In some countries astrological predictions are used as the basis for taking life's important decisions, and Western man's delegation of the decision-making process to an unseen Expert figure is the same mental shelving of responsibility. It is like tossing a coin or flicking a dice, only in this case of Experts they use character tests or handwriting analysis as they arbitrate over appointments (vocational guidance is a classic case in point).

The intensity of interest with which these procedures are followed is clearly evident in newspapers and magazines, where the same kind of test approach is avidly obeyed. Thus the Sunday newspapers may feature a quiz headed:

HOW GOOD A PARENT ARE YOU?

or perhaps a women's magazine will proclaim:

They are essentially the same as any other Expert advisory system in the following respects:

a) they are composed at random by Experts who invent the questions, and decide which are the right answers;
b) they lack any feedback, which helps to make them consistently meaningless.

As a rule it is possible to work out which of the various options is the right one, so that everyone who does the quiz ends up with a high (i.e. favourable) score. You cannot get a 100% rating, though. The Expert is the only person who can do that, since he or she sets the quiz and always moulds it on personal preferences.

Maximizing your Chances in School
What society needs, like all successful biological systems, is cybernetics – feedback. Feedback involves the application of the output to regulate the input, rather in the manner of a governor on a steam-engine. Too much speed causes the governor to turn too fast, cutting off the steam supply until the rate settles back to normal again. The thermostat is another familiar example. But the Expert knows that feedback would undermine the very foundations of Nonscience.

Feedback regulation is found throughout the ecological systems of our planet, and the whole functioning of any living organization depends upon it. Some principal systems that lack feedback are:

CIVILIZATION
MADNESS and
CANCER.

Oh – and Nonscience, of course.

From the earliest stages of training it is easy to see how the feedback loop is eliminated. Take teachers. Are there any rules about how they are selected or their ability actually to teach anything? Is there any feedback in the way teachers behave? There is

91

generally no real appraisal of teachers, so that a poor teacher might be prevented from wreaking havoc amongst the tender minds in their care. In substantiation is the estimate of 15,000 poor teachers in Britain, published in May 1981 (and estimated by teachers, incidentally). There has been much discussion in recent years about whether teachers should keep records of pupils' behaviour and abilities, just in case this smacks of elitism. Far more interesting would be the keeping of records on the teachers by the pupils, which would soon show up an individual who was consistently humorless, muddle-headed, sadistic or just plain wrong. No wonder Experts shy away from that idea. Meanwhile, if students fail their examinations (of which more anon) then it follows that it is always the pupils' fault and never the teacher's.

> Meaningful dialogue between a student and his facilitator (formerly known as a teacher) will hopefully promote a life style showing greater commitment to today's values. Through insightful conferencing the student will be able to conceptualize his future place in modern society and will therefore have options previously unavailable in an academically oriented classroom Present learning experiences as presented by our faculty now provide in-depth studies that will thrust through out-dated socio-economic barriers and inservice meetings will further allow the structuring of programs utilizing space organization and team teaching.

Naturally, one teacher knows little about a given pupil compared with how much a pupil has experience of the teacher.

Suppose a teacher takes a class for one hour twice a week for thirty weeks. And let us suppose there are 25 pupils in the class. The number of hours contact on an individual basis which the teacher can have is therefore:

$$\frac{2 \times 1 \times 30}{25} = 2 \text{ hours } 24 \text{ minutes}$$

Conversely, each pupil has a full ration of knowing what it is like to be on the *receiving* end, i.e. 60 hours. In consequence it can be calculated that the pupil in

92

this case has had *twenty-five times* as much exposure to the teacher's ability to teach as vice versa. Since there are more than two dozen pupils to each teacher, then the cumulative amount of exposure-hours is 25^2 hours $= 26.04166$ days.

This proves, beyond the shadow of a doubt, that such a class would be 26.04166 times more experienced in how good the teacher was, than the teacher would be about the pupils. But of course, it would be argued that pupils are not so good at judging these matters as are teachers. Very well, let us say that a teacher's ability to judge pupils is 350% that of the pupils' accuracy at assessing teachers. (I doubt whether that is the case, but let us accept it for the sake of the argument.) Then we conclude that the final score in our hypothetical class situation – I nearly said class *room* there by mistake– shows that pupils are 7.440476188 times better at judging their teachers, than the other way round. On that basis, they ought to have a voice.

But is this kind of reciprocity compatible with the aims of Experts? Perish the thought!

How to Train as an Expert
All would-be Experts should observe these *Pupillary Principles*.
a) *Conform to what the teacher says.*
If a class is taught some theory that is arbitrary, wrong or merely out of date, then take no notice. Do just what the teacher says, and on no account argue. That would make you seem a probing, insightful type, not suitable for training as an Expert under any circumstances.
b) *Do not be intellectual.*
Any attempt at bringing in signs of intelligence, or trying to reveal intellectual traits of an old-fashioned sort, smacks of being 'smart-alec'. Do not do it: you will be marked out as an enemy of the system. (Note: As well as that, any Nonscience-orientated teacher will not understand a word you say anyhow.)
c) *The Rapt Attention Mode.*
It is a comparatively easy matter to learn how to sit

93

upright, looking bright and attentive, whilst actually you are miles away thinking of something else. This is a vital attribute for boring lectures, which Experts invariably give. It is equally handy for discussion groups with the long-winded, and will eventually help when you visit symposia and are stuck with a professional Expert rabbiting away for hours at a time. It is important to remember to nod from time to time in fervent agreement with whatever is being said. In this way you avoid being asked any questions that would embarrass you, since it appears that you would only answer in lengthy and effusive detail which would show you too know more than the boss.

Should you wish to go to sleep, make sure you are propped up securely in a corner with a book open on your lap, as though following with phrenetic interest every fact that is being put forward, every cliché that you know is being trotted out. (I make no excuse for emphasising the need for a sound state of equilibrium in this mode. I once fell headlong down the aisle in a particularly steep lecture theatre during a horrendously boring monologue, having gone to sleep in the manner prescribed whilst propped up in the end seat of a row. Though well ensconced it is possible I was nudged warningly by a neighbouring student since I was allegedly muttering about salacious experiences and snoring at the time, so that might explain my downfall.)

In short, always look as though – in some way – you are paying fullest attention. Above all, remember it is most important not to yawn. If the impulse is irresistible, then convert it into an expression of open-mouthed admiration instead.

Your turn to bore, rather than be bored, will come.

d) *Be sympathetic.*

Work out the teacher's opinions so that you can be ready to agree with them, no matter what they are. When you do agree, agree fulsomely. His own bents will soon emerge, and you must remember to concur with each and every one of them. You may find it helpful to join with the teacher's choice of fashionable

94

clothing, or drop in snippets about the superiority of his own make of car, for instance. That can suggest subservience and admiration to the point of hero-worship.

If by chance you are put on the spot, resist any temptation to cross swords with the teacher on an issue he holds dear. Just *lie*, lie like hell.

e) *Laugh.*

f) *At the right moment.*

Remember that a teacher is giving out a set series of notes and he inevitably covers the same facts year in, year out, no matter what happens to the framework they hang on. You will hear his regular jokes trotted out monotonously, and his flat delivery will immediately betray his own familiarity with them. So if he says: 'Copper nitrate – oh yes, copper nitrate, anyone know what *that* means?' do not groan, do not even allow your face to betray that you knew the answer all along. Wait for him to reply: 'It's what the police get paid for doing overtime!' and laugh, laugh, laugh.

Examining Examinations

At the end of the course you run into the examinations. These are random memory-tests which take place on a hot and humid summer afternoon designed to prove that you can sit still and do exactly what you are told no matter how trying the conditions. Examinations are not intended to show how clever you are, since if you were clever you would hardly be wasting your time sitting indoors under such conditions when there were better things to do.

Sometimes people complain of 'examination nerves', but they deserve to fail anyway. That sort of person is likely to chicken out just when an exciting experiment gets dangerous, or when doubts begin to arise over some interesting new ban.

Sometimes 'examinations' have been withdrawn as part of progress, and substituted by 'assessments'. These are the examinations under a different name: they produce scores built up in much the same way, and by exactly the same people. People who rebel against 'examinations' are usually quite taken in by the same procedure when named 'assessments'.

In an examination situation, there are two basic types of question that you may come across.

THE FIRST concern the pet hates of the examiner. Thus the question will demand a discussion, a comparison or a discourse on the merits or otherwise of a couple of propositions. What is required here is a careful, point-by-point reiteration of exactly what the teacher taught you on that topic. You may have meanwhile encountered a rival theory advanced by someone else, indeed that might be in your opinion a vast improvement on what you were taught in the first place. But it is only a drop-out and a non-Expert who says so. If you insist on bringing it up, then do so in an air of firm disapproval.

THE SECOND category of written question covers factual material that will have been learned by the student from the text book or lecture. They are simply a restatement of data.

Lectures are based on the principle of reading out, in a muffled monotone, passages from books that could as easily be read by students. Some people even ask whether this is just as good an alternative. Shouldn't students get on with enjoying life, leave out half those classes since it is all in the book anyway, and just sit the examination when they had learnt it all? This misses the point. Experts know perfectly well that you have to make people obey. Not only that, but there is a high chance that the pupils like to have it all read out for them like mother reading bed-time stories in infancy. Furthermore, the way some words are pronounced changes from place to place, and fashions come and go at a machine-gun pace (recent innovations have been psychiatry pronounced as *sick'iatry* and microbiology as *mick'robiology* which, until we ride *bissicles* and look down *mickroscopes,* I shall ignore).

The error that many people make was well illustrated by the exasperated lecturer who came across a student making some original and somewhat unorthodox measurements of muscles and surface body indicators in an anatomy course. 'You are suppose to learn the *course*,' he fumed, 'not the *body*!'

The Practical Examination

These are tests of instruction-following routines. They are one way of demonstrating how the monotony of method has been instilled into the mind over the years. There is no merit in trying to short-circuit the procedure. Often a quicker and more efficient scientific approach to the problem will arise, but that is not what you are being tested on.

Let us take a practical examination from the past as an illustration of how *not* to go about it. (Incidentally, the author would take a pretty dim view of anyone unkind enough to associate his name with this best-forgotten episode!) The student was sitting chemistry, and part of the exam involved identifying a metal sample. The instructions were to use the equipment available in the laboratory to identify the metal in the object provided.

The candidate was given his unknown sample by the examiner, who laid it down gently on the bench. That was a mistake from the start, for the bench had a slight slope and the spherical object rolled silently along the bench and then, with a resounding clunk, it fell into the sink, chipping the enamel as it did so. The metallic sphere then trundled across the bottom of the sink and disappeared down the deep plug-hole.

With a penetrating glance in the student's direction, the examiner produced a second example from his brief-case, and placed that – with heavy emphasis – in front of the student. To prevent it rolling away again, and with an entirely justified look of condescension, the examiner grasped a tripod that had been standing nearby with his finger and thumb, and neatly flicked it over so that the ball was trapped inside the enclosed space.

Unfortunately the examiner had not noticed that the tripod had been standing over a bunsen burner for half an hour, and the top of it was already at a dull red-heat. As a result – after a pause of a few seconds – it began to brand its way into the bench, emitting crackling flames and sparks, and sending out fierce jets of pale brown fumes which quickly filled the room. The coughing students followed the examiner (who was by now perspiring freely) out of the room until the air cleared.

The candidate, after dousing the flames with a beaker of water, began his analysis. He need hardly have bothered, of course, for although the students were now filtering back into the laboratory there was an air of hilarity about the whole proceedings. The first thing the candidate noted was the appearance of the ball-bearing he had been given. It was rusty. Only one metal rusts. Immediately he realized that the metal must be iron.

So he asks the examiner: 'I wonder if you could provide me with a magnet?' There is a change in the atmosphere. Through tightly gritted teeth, the examiner repeats: 'A *what*?' 'A magnet, please,' asks the student again. There is a pause, as the examiner inhales deeply through his nostrils. 'A – *magnet*,' he proclaims. And after a few minutes, rummaging in dark and dusty cupboards in the corners of the laboratory, he produces what he has been asked for.

Eagerly the student lays the magnet on the bench. The ball-bearing rolls towards it and with a satisfying clanging sound it clings firmly to the magnet; indeed it proves to be difficult to prise it free. The result is clear. The metal sphere is rusty, and iron is the only metal that rusts. It is also strongly attracted to the magnet, and only one metal – iron, of course – behaves like that. In two simple moves the identity of the element has been proved.

You could argue that since the examination rubric had commanded: *IDENTIFY THE SAMPLE OF PURE METAL PROVIDED, USING THE EQUIPMENT AVAILABLE IN THE LABORATORY*, the student should get full marks. Come now; this was not a test of problem-solving, but of problem-generation. He got 0% for that question, and quite right too.

Marks

What matters is that you get marks. It rarely bothers people where you get them from. In one recent example,* school-children were asked to mark examinations sat by 11-year-olds, which provoked a deal of controversy particularly when it emerged that clerks and local government administrators – instead of qualified examiners – had also been marking papers in their spare moments.

But in 1978 a report** prepared under Sir James Waddell pointed out that the examination results obtained by school-leavers were – in essence – irrelevant. The London *Sunday Times* explained:

The remarkable thing about O-levels and CSE is that they convey so little information about the children who take them. A grade three in English, for example, says nothing about a child's ability to spell, punctuate, or write clearly and legibly ... Similarly, an exam grade in maths gives no information about a child's particular mathematical skill ...

But does anyone take notice of the clear and uncontested irrelevance of examination marks? It appears not. I do not know of any research which suggests that examinations might actually give a useable index of ability, whereas there is no lack of information that proves the opposite. However it has become a grand tradition that – although you may reach such conclusions – you must not actually implement any of them.

The definitive example was the survey by Philip Hartog and E. C. Rhodes. They took 15 papers that had all been given exactly the same percentage mark in an examination, and sent them off to other examiners for re-assessment. When the papers came back the second time the highest mark they had

* 11-plus Marking by 6th form angers Teachers, *Daily Telegraph*, 22 January 1979
** Waddell report [in] Wilby, P., Why exams need examining, *Sunday Times*, 16 July 1978.

gained was 71%, whilst the lowest rated only 21%. In two noteworthy examples, a paper placed at the top of one examiner's mark list came at the *bottom* of another's.

More recently a group of candidates sat papers in the same subject, but set by different authorities. According to one of them, 27 passed whilst only one failed; but another authority passed only 3 of the candidates and failed 25. And then there was the case of the batch of students who all failed their examinations. Their marks were queried and after investigation it transpired that they had all actually passed, instead.

You might think that the results of Hartog and Rhodes would have caused a sensation and led to a radical revision of our examination system; or perhaps it occurs to you that there has not yet been time to change things, and that a revision is due any time now. Frankly, I doubt that. Their results were published in 1936. We are bedevilled by examinations now in exactly the same way as we were then.

There is only one difference, and that is our results are poorer, and they mean less now, than at any time in the past.

CHAPTER SIX

Futurism

Planning the future of the training process takes us into dangerous ground, partly because we are very much in the lap of fashion, and partly because Experts tend to shy away from what people expect. Whatever else happens, the future of schools and training institutes.will centre on ceaseless change.

If a school has not altered recently, change it now

The Expert's capacity to produce random alterations when least expected should be brought to bear on the school-situation situation. If the school used to be small, enlarge it; if it was merged a few years back, change it into separate units again. Rename it. Alter its catchment area. Replace the 'head' (preferably with someone less smart than the present incumbent).

Keep the syllabus turning over

What is taught in the school must be presented differently just when people least expect it. There are ways of shocking people into a frightened fervour of obedience if you give subjects different names, teach them in a different order, take out favourite topics and slide in something inexplicable in their place. Try never to leave pupils feeling settled, or the fun of Nonscience begins to wane. It is like musical chairs.

Abandon classes

This fundamental gambit has had a wide popularity amongst Experts with a *penchant* for causing the greatest havoc. It produces a great sense of power and innovation. The class itself is abandoned, and children are allowed to wander from place to place at will, choosing whatever they like to do at the time, whenever they want to do it. This causes such a chaotic upheaval in the day's routine that everyone is too preoccupied to worry about any external matters.

Abandon the class room
In some schools the rooms themselves have gone, to be replaced by open-plan screened units between which the bemused children wander. Each class can be overheard by others, and the respect for the lofty Expert who came up with the idea in the first place is enormously increased into the bargain.

Tape recorders, amplifiers and deck-top computers all serve to give the teacher's function an auro of complexity and professionalism. Experts approve of the distancing effect on pupils that results.

Abolish lessons
This, I am sure, will come next. The teacher/taught interface situation can be maximized with incremental viability in a non-structured optimization rationale of data-acquisition input modalities where time-dependent communicational norms are subverted through electional discipline-identification at psychologically referred impulse stimuli, as an Expert explained to me only the other afternoon.

102

Replace 'old' subjects with the 'new'

Readers will know of many subjects which have hit schools over recent years, The New Maths being the most familiar example. We will look at that in a moment, but what new subjects might we expect to see in the coming months?

THE NEW ART in which visualization is eliminated. Students will be allowed to express themselves through non-art media, so that no drawing, painting, drama or art appreciation is included. Indeed pupils who can actually produce pictures should be discouraged from this, as it shows a representational mind which the new art would decry as revanchist and founded on elitism.

THE NEW ENGLISH abandons spelling, syntax, vocabulary, but encourages children to develop a deep-seated indifference to literature at all levels. English is, in this way, made to seem apart from the every-day world, and presented as something boring, abstruse, esoteric. Significant progress has already been made, of course, but future classes in The New English (or non-classes, if that is how the fashion moves) will really set illiteracy on the map.

THE NEW SEX is beginning to make its mark, too. Here, the end we wish to attain is the degradation of sexual activity into a physical act of some repugnance. It seems that the pleasures of falling in love and love-play are diverting too much attention away from the interests of Expertism. So a form of sex training has been introduced which concentrates the mind away from these delicate pleasures, and draws a distinction between that form of time-wasting and mechanistic intercourse as a short-term fulfilment (short-term orientated fulfilment situation, I mean). The concept of cerebral orgasm and its exquisitely pleasurable exploitation is done away with: penises are 'inserted' and intercourse 'takes place'.

The leaders of thought in this enterprising new field look with pride on the first generation of young people who have been largely desensitized against sexuality and who – failing to find personal relationships in the least satisfying – are willing to turn to Experts for leadership instead. And, at the same time, large amounts of investigation into the sexuality of children are being recovered from the dusty pages of the journals:

The girl jumping a rope acts out the to-and-fro movement of the man during sexual intercourse. Her own body takes the part of the active man, while the swinging rope imitates her own body adjusting to the movement of the man's. In this game, the girl acts both the rôle of man and woman. Thus the girls go through unconscious preparation for their future sexual function as women.*

I'll wager you did not know *that* before!

THE NEW MATHS remains the example of modern training that everyone should admire. In principle, what it has done is to down-grade mathematical attainments until children are unable to work out the simplest sum. Critics who point this out are immediately told: 'You cannot make assertions about a decline in standards, because of the change in mathematical content ...'

One form of comparison is to present today's schoolchildren with an examination from an earlier era, and in 1980 a test was organized by the National Foundation for Education Research,** in which a sample of bright 11-year-old children were given a set of question from an 11-plus examination set in 1937.

In each case, the old-fashioned measurements of dimension, money, etc., were translated into their modern metric and decimal (or centimal) equivalents. Some of them are as follows:

* Sonneberg, M., Girls Jumping Rope, *Psychoanalysis, 3:* 57-62 (1955).
** Maths tests [in] Wilby, P., The Modern Way to fail the 11-plus, *Sunday Times,* 25 May 1980; *ibid,* Teenagers who can't count, 3 Feb 1980; and Maths experts 'get school sums wrong', 25 June 1978.

1937 Questions

1: Add one-half of $5\frac{3}{4}$ to 4 7/12.

2: The school gate opens at 8.50 a.m. and closes at 3.30 p.m. Play-time in the morning lasts from 10.45 a.m. to 11.00 a.m., and the lunch-time is from 12.15 p.m. to 1.30 p.m. How many minutes are there for lessons in the school day?

3: Mary has got £1.70 in her savings and John has £1.42. If John saves 9p per week and Mary saves 5p per week, how many weeks will it be before John has more money saved than Mary?

4: Mrs Smith, Mrs Brown and Mrs Jones go for a holiday together. Mrs Smith pays all the railway fares of 80p each. Mrs Jones pays the entertainments which come to £1.65 altogether, and Mrs Brown pays the boarding-house bill of £2.35 each. At the end of the week they all settle up, the two other women paying Mrs Brown something to share the cost equally. How much does Mrs Smith pay Mrs Brown?

5: A wall is 10 metres high and 27 metres long. How much will it cost to paint one side of it if the paint works out at 6p per square metre?

1980 Questions (Modern Maths)

A: A four-centimetre square has been cut into five pieces. They are labelled A, B, C, D, E. The shape of D is square. Which two pieces can be fitted together to make a square?

B: Which fraction is the smallest?

$\frac{1}{2}$ $\frac{3}{4}$ $\frac{3}{8}$ $\frac{1}{4}$ $\frac{5}{8}$ $\frac{7}{8}$

C: Here is part of a railway timetable. It shows the time trains leave Paddington and the times they leave stations on the way to Bristol.

PADDINGTON	06.15	07.50	09.45	10.45	11.45	12.45
READING	06.55	08.24	10.19		12.19	
SWINDON	07.38	09.08	10.54	11.19	13.04	13.19
BATH	08.14	09.44	11.29	12.18	13.42	14.22
BRISTOL	08.30	10.00	11.45	12.35	14.00	14.40

Using the information in the timetable, which is the

105

latest train you could take from Reading, to get to
Bath before 2 o'clock in the afternoon?
D: A graph can be used to convert miles to kilometres, or kilometres to miles. Use the graph to
convert 1.6 kilometres to miles.

ANSWERS:

1: 7 11/24	A: B and E
2: 310 minutes.	B: one-quarter, $\frac{1}{4}$
3: 8 weeks	C: 12.19 p.m.
4: £1.30	D: 1 mile.
5: £16.20	

The findings, when the papers were all marked, was
that the vast majority of today's 11-year-old schoolchildren would fail the test outright. The earlier test
was devised to select children for the grammar
schools, and on that basis only a handful of the
children tested in 1980 would have got a place. Marks
were given out of a total of 100, and the 1937 children
who sat the test (only a part of which is shown here)
obtained an average score of 55%. Today's children,
by contrast, averaged only 12%, a catastrophic drop.
In defence of the results, a headmaster who was
involved in the research stated that the newer test
involves a broader range of concepts and includes
items – such as the timetable question – absent from
its 1937 counterpart.

This appears to be irrelevant. If the concepts in
today's paper are covering a broader ground, then the
narrower limits of the 1937 paper should be even
easier. In that case, the 1980 children should have had
a higher score, and would doubtless have found the
narrow-minded 1937 paper exceedingly easy. Secondly, the same kind of extractive mental process is
indeed found in the earlier paper. The question about
the number of minutes available for school lessons
(Question 2) utilizes much the same conceptual basis,
and any child who could have done Question 2 in
1937 would have found Question C in the 1980 paper
elementary.

This betrays the essential difference between the

two papers: in 1937 the pupils had to work out the answer using a great deal of common sense and a considerable amount of effort. The 1980 version does not demand anything like as much work. Most of the questions you answer simply by looking at the question, and then writing the solution straight down. In 1937, for instance, we have Question 1 posing a complex calculation utilizing fractions. The 1980 Question B on fractions demands only that you write down which one is the smallest.

Similarly, the 1937 Question 5 on area involves a systematic calculation on costing per unit covered; the 1980 Question A asks only that you spot which bits make up a square. The approach demands far less psychological application and only the minutest scrap of effort; it is more like something out of a Christmas cracker. In essence, this is what has happened – the maths examination has changed from being a severe test of ability to being an infants' game, a riddle, which anyone can pass and which you hardly have to lift your pen to accomplish successfully.

Quite how much damage this does is hard to assess, but it is unlikely that any real indication of the slow but insistent lowering of standards will ever be obtained. In 1978 there was a sign that the British government had been influenced to start gathering data on this important issue: a series of tests of mathematical ability graded on a scale of relative difficulty was given to 12,000 schoolchildren as the first stage of a monitoring process. The tests were devised by a Danish professor of statistics, Georg Rasch, and he claims that they have an inbuilt flexibility. Thus, if a new form of mathematical calculation becomes the rage, then you can substitute the new one for its predecessor without upsetting the sequence. 'All you have to do is make sure they are of the same degree of complexity, so that the new one is as difficult as the old,' I was told.

Needless to say, just as one bunch of statisticians were insisting that the Rasch tests were a wonderful idea, there was an equally vociferous group claiming the exact opposite. 'It is absurd,' proclaimed a professor at London University; 'some children may be good at algebra, say, and bad at geometry.' In the eyes of the critics the new system would produce endless confusion, and the results would be entirely useless for any serious research.

Meanwhile, which side is right? The answer is neither. That is the joy of the Nonscientific system when it really gets going: people capitalize out of the conflict, and Experts on both sides clash with the occasional frustrated scientist, who is trapped in there, looking for the truth. And all the while the money keeps coming in. The initial grant for this work was £470,000 which is enough to keep the conflict reasonably fuelled for half a decade at least.

Here are some examples from the Oxford Examination of the first-ever School Leaving Examination.

JUNIOR

Qu 9: In a screw press, the screw has 4 threads per inch, the power is applied at a distance of 14 inches from the axis of the screw, and the surface pressed is 110 square inches. Find what power must be applied to produce a pressure of 1 lb to the square inch.

Qu 13: A bow is stretched until the tension of the string just equals the pulling force. What is the angle between the two parts of the string?

SENIOR

Qu 3: What is said to be the velocity of light?

How was it first deduced by Römer, and subsequently confirmed by Bradley?

Explain fully their observations, and the deductions from these observations.

How far do they show the similarity of the nature of light when derived from different sources? Do you know in what countries, and at what date Römer and Bradley lived?

Do all kinds of sound travel with equal velocities?
Prove your answer, and mention any facts you know as to the velocity with which they transverse different media.
How did Wheatstone try to deduce the velocity of electricity?
How much did his experiment really prove?
What circumstances practically affect the velocity of electricity, and at what rate is it found to travel in practice?
Does its velocity depend upon the intensity of the battery from which it is produced?
Compare the laws that seem to regulate the velocities of light, sound and electricity.

Experts delight in telling us how far modern education has developed. These sample questions from School Leaving Examinations in 1858 show how educational standards in that far-off, unenlightened age compare with our modern era of scientific understanding.

In case you are interested to know what kind of performance you can detect in the modern British schoolchild sitting a mathematics test, then the first indications were published in January 1981.* They showed that half of British children (metrication or no), did not know that 1000 grams were the same as 1 kilogram, nor did they know how to divide 8 into 816, nor could they undertake subtraction sums where 'borrowing' from the next column was involved.

Indeed, the findings as a whole proved that a very large number of children are unable to perform simple calculations. The questions were like these, which are shown in order of complexity, easiest ones first:

1: Find which number stands for V:
$$12 - V = 8$$

* Assessment of Performance Unit, report on modern maths teaching, London, 1981 (see *Sunday Times*, 25 Nov 1980).

2: Put these decimals in order, smallest first:
   0.3, 0.1, 0.7, 0.6
3: 6/10 + 3/10 =
4: 256 children are going to have tea at Christmas. Eight children will sit at each party table. How many tables will be needed?
5: A man can cycle a mile in 5 minutes and walk a mile in 20 minutes. How much time does he save when he cycles the 3 miles to work, instead of walking?
6: 381 x 11 =
7: What number is 10 times 0.5?
8: 1/6 + 2/3 =

The answers are given below, along with the percentage of children in the test who got the answer right.

QUESTION	ANSWER	PERCENTAGE WITH RIGHT ANSWER
1:	4	88%
2:	0.1, 0.3, 0.6, 0.7	79%
3:	9/10	69%
4:	32 tables	53%
5:	45 minutes	51%
6:	4191	49%
7:	5	34%
8:	5/6	28%

It was universally agreed that the results, when they were published, showed an astounding lack of ability in the modern school-child. Clearly the effects are biting deep: when the results of the survey were leaked by the *Sunday Times* two months before official publication date, they proudly gave the answers in a fashion that would make any Expert proud.

'What number is 10 times 0.5?' they asked, and then triumphantly printed the answer:
   '30°,' it said.

And what of children who are older? The figures obtained in the Assessment of Performance survey for 15-year-olds throw some light on that. For example, it

110

showed that 40% of 11-year-olds were unable to multiply 76 x 7. After another four years at school the figure had changed. The percentage who could not do that sum was by then 20%.

Who would have predicted the day when, in the midst of talk of technology and the wonders of the space age, people of 15 could not work out what 76 x 7 amounts to? Well, if Experts have their way, we may yet attain the ultimate goal where young people do not even know what '76 x 7' *means*, let alone what it makes.

To give you a taste of the extent of the problem, here are three of the questions set to the 15-year-olds:

1: $1.8 \div 3 = ?$
2: The annual income of a corporation increased by 70% during the last year. What is the ratio of its present income to its income a year ago?
1/70 7/10 10/7 17/10
3: A man left London at 10.15 a.m. and arrived in Manchester 190 miles away at 3 p.m. What was the average speed of his journey?
4: Write a number in the space to complete this statement:
   $73.45 = 70 + 3 + 0.4 + ?$

Many of the youngsters who completed the test found the questions difficult. The question of ratio of incomes (i.e. what is the ratio of 170% to 100%/ Question 2) was wrongly answered by 85% of the candidates sampled. And 82% could not work out the simple question of average speed in Question 3. The full answers are given below, along with the percentage of 15-year-olds in each case who gave the right answers:

QUESTION	ANSWER	PERCENTAGE WITH RIGHT ANSWER
1:	0.6	66%
2:	17/10	15%
3:	40 m.p.h.	18%
4:	0.05	51%

The questions were not in any way complicated; they fit the kind of demands that you might expect a reasonably educated 15-year-old to tackle with ease. Yet the modern child fails to complete them accurately – why, one-third of the sample could not even divide 1.8 into three.

It would be wrong to imagine that this necessarily means that the children are unintelligent. Often it is a failure on the part of adults to teach children the basic skills they need to acquire. We like to imagine that youngsters acquire mental ability as they get older, but that is only part of the truth. In every case there is a profound need for the latent skill to be developed, for the child to be educated, drawn out and encouraged. It is this which today's teaching does not do.

As a fine example of the extent to which even simple logical argument and basic deduction can be strangled, I may cite the following problem set to the children in the test for 15-year-olds. There were two statements:

a) All pupils at Castle School were boys
b) K. Smith was a pupil at Castle School.
The question was: does this mean K. Smith is a boy or a girl? In the full spirit of Nonscience, 10% got even that answer wrong.

Testing tomorrow
The micro-chip mentality will in time put an end to the traditional training and selection of young people. Note that I do not say 'the micro-chip will put an end' to it, but that the 'micro-chip *mentality* will'. Micro-chips are cheap, and they are tiny, but they do not suddenly make automation or robot control or memory banks or computer units possible for the first time. The micro-chip mentality, on the other hand, makes people believe that micro-chips are utterly miraculous, and that they can achieve anything, and streamline almost every process known to man (along with a lot that aren't).

112

This is the mentality which will affect the training process in the coming years. To begin with, there will be automated training machines, somewhat like the language laboratories already in use. The individual will sit at a desk covered with twinkling lights. Reedy computer voices will squeak into headphones. Instead of writing implements there will be a kind of electronic typewriter which will duplicate the writing process with two differences from the conventional way of doing things:

i: it will be infinitely more complex and
ii: it will eliminate any great need to use the brain.

Instead of morning register, with students answering 'here' or 'present' when the teacher calls their name, a press-button will signal:

PUPILLARY MODE IN DATA-TRANSFER
OPERATIONAL INTERFACE, BLEEP, BUZZ

and the day will then start.

As this high-tech training conveyor-belt system hots up, more and more young people can be force-fed with the raw facts they need unblinkingly to obey their masters. It cannot be long before we enter the ultimate examination situation, which I have extrapolated from the exam papers already cited to give a prediction of what we will have by the end of the century:

2001 EXAMINATION PAPER (PREDICTED BY EXTRAPOLATION MODEL)
1: How do you spell MODULE?
2: Which figure is the biggest, 2 or 2?
3: Can you explain the difference between an ancillary reciprocal facility and a bunch of daffodils?
4: If I invite 12 people to a party and all but 3 attend, how many chairs will they need?
5: What is $22 + 35.4 \times 14\frac{1}{2} + 99$?

ANSWERS:
1: 'With difficulty' is the correct answer here.

2: 2.
3: The correct response is 'no'.
4: One each.
5: $22 + 35.4 \times 14\frac{1}{2} + 99$ is a *sum*.

From here it is only a short move to the mass-production of qualified youngsters on a batch-sampling basis. There would be no need to test each one. The training procedure would be then so standardized, so perfectly regulated and so utterly, painfully predictable, that all you would have to do would be take one from each batch (like a tub of yoghourt or a bottle of lemonade) and analyse him or her in detail. Our tendency to value human life is already plummetting, so there should be no difficulty in analysing the brain of the selected individual. It could be scooped out and given the full treatment in a chemical analyser, and the details of attainment in the class laboratory situation carefully integrated to produce a fool-proof analysis for the whole batch.

The result would be the production of entirely mindless and predictable young students of uniform abilities and standardized profiles. Sometimes I think we are more than half-way there, already.

CHAPTER SEVEN

College Days

The purpose of a college or a university is to provide a closetted and confined period of indoctrination for malleable minds. At the end of the course, the typical student graduates with the conviction that he or she knows pretty well everything there *is* to know about their particular subject. In point of fact, not only do they know next to nothing, but their capacity to realize this has been subtly extirpated by the course of training they have had to endure.

The modern university complex is constructed like a prison camp. There are crowded concrete living quarters for the inmates, in which a furnace heat is circulated through the rooms and out through the windows, which are always kept open to allow the heat to escape. For this reason the climate around a university is quite different from what you find in the nearby town. Spring comes early in those heat-liberating parts, the snows melt long before they should, and the fevered brow of the student is well-tuned to the task in hand.

MOST STUDENTS have their schoolday dreams shattered once they get to college, says a research report today. Three out of five feel let down by teachers or disappointed by the quality of life on the campus. And it is this unhappiness which leads to unrest and disruption declares the report's author.

The campus has bars, snooker rooms, squash courts, cinemas and libraries; all intended to keep the student occupied and away from the pressures of the real world out there. There may even be a closed-circuit television system or a private radio station to add to the claustrophobia and introspection.

Staff are not allowed to come into intimate contact with the inmates, however. There may be a house-master figure, who lives in a dingy modern building,

115

as a pervading presence in case of trouble, but staff otherwise keep away. It is a bad point to become too close to the very people who are supposed to be building up a level of respect for their superior. Much of the week is taken up with lectures, the recital of scraps from whatever koran is in vogue at the time. There are also tutorials, at which the staff member expounds exactly what he wants the students to believe, and seminars, at which the inmates recite back what they were taught to believe in the tutorials.

It used to be said that universities were the seat of learning. That is now an out-dated and unfashionable premise; the new ones are seats of organizational strength and little else, and many of the oldest and most reputable colleges are heading the same way. People once thought that universities were a ferment of political activity, but the creed of Fashionism has overtaken that. Now, the only views that can be preached are those that are fashionable. Totalitarianism, for example, is welcome as a subject for meetings as long as the speakers proclaim their alliegances to the political left. The same dogma extolled by anyone with rightist views is unfashionable, and therefore such speakers are simply kept out, or if they do sneak in by some subterfuge, they are barracked and expelled or merely attacked on site.

In this manner, any political activity is kept strictly within bounds. You could draw up a list of permitted subjects for discussion which paralleled the pop chart for topics which we saw in Chapter Three. In recent years the charts were headed by legalizing drugs from marijuana to LSD, followed closely by campaigns for women to become miners, truck-drivers, lighthouse keepers and obsessional executives; and then by gay lib, which was almost superseded by child pornography.

Today's students are astonishingly racist, too. Whites dislike blacks and Asians with a rare fervour, and they distrust Arabs and orientals just as much. The Asians nurture dislike of each other based on caste, the Arabs on religion or political persuasion, and the Japanese on family wealth. There are very

116

many senior students in medicine, for instance, or dentistry, who say openly that they will under no circumstances treat 'one of those black bastards' or 'any Asian peasant'. What an unpleasant individual the white Expert is.

Naturally, there are mechanisms to disguise all this. Courses are organized and societies set up to insist that racial harmony is the order of the day. Slogans which say things like: SOLIDARITY WITH OUR BLACK BROTHERS are exhibited in the strongholds of racial prejudice. These moves are always organized by whites, of course.

Foreigners are disliked by staff, too, but they are taken on because of all the money they bring in (they are now charged a high fee for attendance, which is regarded as being one way of recouping some of the foreign currency losses through aid and oil revenues to OPEC). However there are some subjects where places are few, and where pressure is sufficient to keep the department well stocked with adequately funded white students. I can call to mind one subject in which there is only one black student in the length and breadth of the nation, the others having been 'turned down' on the ground of 'interview performance', and we all know what *that* means.

Today's students are distinguished by their astonishing similarity. They have been trained to wear the same dress, moving on from the baby-gro's of infancy through the dungarees of childhood to the student standard garb. This compulsory uniform is jeans and a sweatshirt. It is worn by both sexes. In the interests of uniformity, students are not encouraged to differentiate between 'male' and 'female' gender of their species. In turn, this leads to a lowered incidence of sexual reproduction. The birth-rates reflect this trend.

The eventual aim of this intersexual uniformity, bland monotony of outlook and identical forms of dress is a homogeneous and predictable student population. I have a feeling that cloning is going to be the ultimate fate of the production of new recruits by the end of the century. That will make the predictable uniformity of students even more reliable, and it will remove any chance of the occasional throw-back who might start to ask embarrassing questions. Though the prospect may seem disheartening, it should not cause too many practical difficulties at a social level: students are so alike already, that the day when they become completely identical will make surprisingly little difference.

The topics embraced by the training procedure include set opinions and received facts. If the student is asked what his opinion on some development might be, it is imperative that he quotes the opinion that he has been taught.

But it is even more interesting to consider what students are *not* taught. They receive little instruction in how to do research, how to write or communicate their ideas (if they ever have any), how to use the journals, how to compile a bibliography, and what to do with one if somebody else has already compiled it. The problems of living – social, psychological, practical – are not covered at all. In this way graduates are well-versed in the limited sphere of factual knowledge in which they have been instructed, and feel confident that they know virtually all there is to know on the subject; while at the same time manifesting an almost total lack of worldliness, experience or wisdom.

Should you happen to be on the receiving end of all this at the moment, it may comfort you to know why the lecturer sounds so boring. He *is* boring. This is largely a consequence of the fact that lecturers are selected because of all the research they are going to do, and not because of the lectures they are going to give. It is also the result of the generally-agreed training that lecturers have to undergo, before they are allowed to lecture: namely, none.

Sitting university examinations relies on the same set of principles as those that pertain in the school situation. Memorize those notes. Follow the rubric exactly. Do just what you are told, and, above all, do not argue or introduce heterodox notions into your answers. At the same time, all students should become conversant with that unique shorthand of examination prose (unique in that – unlike other, more familiar forms of shorthand – it is *longer* than the original). Thus, you would not write 'Don't know' against a question, you would put *Definitive data on this point are unavailable.* The essential phrases that everyone needs are as follows:

CONCEPT	EXAMINESE VERSION
I think some other metal does this too, but I can't for the life of me remember what it's called.	
	Observable phenomena comparable with those here delineated have been reported to occur with other elemental metals described in the literature, q.v.
Mind you, I am not sure that it *always* happens, though.	
	The correlation of these parameters should not be taken as automatically predicative.
What on *earth* do I do next?	
	The procedural techniques of routine investigation are then followed.
I forget the other plant that does it.	
	Related characteristics are found in other species that have been documented by previous workers.
I am much too tired to write any more.	
	Further evidence in support of this view is legion, though clearly beyond the scope of this representative essay.

119

To be perfectly frank I do
not know whether the
statement is right or
wrong.

*The consistency of these experi-
mental data is interesting, since
it lends support to the thesis
advanced; however it would be
erroneous to ascribe any degree of
definitive accuracy to prognostications of such an essentially
pragmatic implication.*

It's all done by something
whose name escapes me
for the moment.

*Further procedures are carried
out utilizing the apparatus/or-
gans/cytoplasmic inclusions
specifically evolved for the
function.*

Students may well be instructed to write a thesis,
and would be advised to consder carefully what this
means in practice. The dictionary defines a thesis as a
dissertation. But it defines a dissertation as a *thesis*
which (even though it is a splendid example of
Nonscience infiltrating the broader pathways of literature) does not help much.

The result you obtain with a thesis depends on who
is going to mark it, and as a rule you know who that is
before you start. By plenty of references to the
examiner's pet likes (e.g. animals as pets) and equally
obvious allusions to what he hates (e.g. – in this
case – animal experimentation) your success is assured. Do not assume that these examples apply to
everyone, of course; there are just as many supervisors
who are staunch vivisectionists, and who are fed
up with animals defaecating on their doorsteps, to
whom the converse expression of sympathies would
apply.

A thesis must be wordy. It must have a lengthy

bibliography (*q.v.*), and it must also be long. If you receive an instruction which reads:

YOUR THESIS SHOULD BE IN THE RANGE 8-10,000 WORDS

then aim for 10,000 and show willing. (Incidentally, it is important not to be tempted to write a thesis 8 words long, explaining to your examiner that technically it is within the prescribed range. Experts would not appreciate the joke.) If you have already discovered how to write bibliographical abbreviations, try not to reveal this fact when you are compiling your list of references at the end. There is a time-honoured tradition of not teaching students how to do this properly, and if you find out by mistake it can look a little too smart if you let on. Rather than writing something like this:

Snodgrass, F. J., Taxation Trends, a Prognostication, *Journ. Internat. Econ., 21* (v): 144-153 (1981)

it would be better to stick to the colloquial formula:

Article by Prof. Snodgrass on 'Taxation Trends', Summer Number of International Economics (facing the Amoco advertisement).

That shows keenness, without betraying any overt signs of ambition, which might be harmful at this stage. The one feature you must remember to include is due reference to the supervisor's or examiner's own publications. Similarly, the important item to avoid, at all costs, is any reference to one of his rivals' publications.

If you play your cards right, manage to associate socially with the examiner (especially if you contrive to act as though you did not realize that's who he was), ask for advice in a polite and subservient manner (and follow it, to the letter), then all you have to do is maintain a cordial and profoundly respectful relationship with your Head of Department, make sure you are dressed suitably when you meet, and a pass is virtually certain. The success rate at the

121

examination level is indicative of the efficiency of the department, and no head wants to create a bad impression on that score.

The Grant

University work is funded by means of grants. For centuries, grants rolled on in a self-perpetuating flood which was never threatened, until 1981-2 when the effects of a flagging economy (itself, paradoxically, the result of too many Experts running too little resources) caused a slowing in the rate of expansion. The outcome was immediate: strong protests, letters and petitions, cries of indignation, which all helped to minimize the damage.

The size of a grant is assessed by a committee, whose members know little about the work, and decisions are taken by people who are easily blinded by a little long-winded gobbledeygook administered in measured amounts. The final arbiter is usually the Head of your Department, anyhow; so keep in his good books if you wish to ensure success. A failure to have a grant application succeed, even when it is the height of fashion, is the invariable hall-mark of having in the past crossed someone who is using this traditional means of indicating displeasure to you.

The smallest grants are awarded by little trusts endowed by philanthropists, whilst the largest ones come from two sources:

a) charities (notably cancer charities) who may well have more money than they know what to do with:

Cancer link with water

BRITAIN'S major cancer and heart charities were heavily criticised yesterday for spending so little of the millions of pounds given them by the public on prevention of disease.

b) governmental agencies such as the EEC, FAO, WHO, NATO and so forth – bodies set up at hyperlegislative level for whom the expenditure of vast sums of money is a sign of power, presige, and purpose.

122

The ecology of grants is itself a fascinating study. There are several organic rules that you ought to know before you make a move.

+ *NEVER IMAGINE THAT YOU CAN SEEM MAGNANIMOUS BY MODERATING A REQUEST FOR A GRANT. IT IS ESSENTIAL TO ASK FOR A VAST SUM.*

+ *NEVER IMAGINE THAT MONEY SAVED FROM THIS YEAR CAN BE USED NEXT YEAR. NO MATTER WHAT YOU SPEND IT ON, IT MUST BE SPENT.*

+ *AVOID FRANKNESS IN A GRANT APPLICATION. HYPERBOLE IS THE ORDER OF THE DAY.*

+ *ENSURE THAT EACH GRANT YOU APPLY FOR IS LARGER THAN WHAT WENT BEFORE. ALL PRODUCTIVE PEOPLE KNOW THAT THIS CREATES THE BEST IMPRESSION.*

+ *CHOOSE YOUR WORDS CAREFULLY. WITH FORETHOUGHT, ALMOST ANYTHING CAN BE OBTAINED WITH MONEY FROM A GRANT INTENDED FOR SOMETHING ENTIRELY DIFFERENT.*

Perhaps the most telling way, in which the importance of the first of those five golden rules can be illustrated, is this account of how *not* to budget. It occurred at a government medical research laboratory where I spent a year in the most junior staff capacity conceivable.

One of the first tasks I was set consisted of drawing up the list of requirements in terms of materials and equipment that would be needed for the year ahead. I did as I was told, and soon put together a shopping list of what we would need. A litre of this, two dozen bottles of that, a yard-and-a-half of something else … you can no doubt imagine the sort of thing.

At last I presented the list to my superior. He grasped my flimsy proposals in his knotted hand and his eyes flicked expertly up and down the columns. As

123

I watched, beads of perspiration began to break out on his upper lip, the bridge of his nose, and then appeared in neat regimented rows over his forehead, collecting in the laughter lines that littered the landscape of his face and gently trickling down in ordered streams like flood water in furrows. Veins started on his neck. He began turning purple. Suddenly, unable to contain himself any longer in a paroxysm of tactfulness, he released the energy in a pent-up bellowed laugh which lives with me to this very day.

I watched him as, with trembling hands, he added the figures up again. It was only then that the full enormity of what I had done began to dawn. The total came, I observed with horror, to over £400 – and all for just one small laboratory!

With some embarrassment I began to mutter apologies, frantically working out where cuts might be made – perhaps a yard would do, or we might manage with only ten bottles – when suddenly he spoke. And then I realized I had not entirely understood his reaction. 'Good heavens, lad,' he expostulated, quivering. 'Four hundred quid for your laboratory? The standard grant is *two and a half thousand*!'

I ventured to explain, searching for justification but only getting myself in deeper as I went, that we could save the £2,100; or at least use it next year. As the perceptive reader will have anticipated already, that merely reduced him to near hysteria.

'If that is all we asked for,' he gasped eventually, 'that's the most we would ever get again. Not only that, but they'd be wondering what the devil we had been doing with all the extra money in past years, if we hadn't actually needed it!'

The first rule, then, is to ask for a huge grant. Overestimate travelling expenses. Never quote 'excursion' air fares; always aim at a full-fare request (first class, even). Claim for first grade hotels, and then sleep in a siding. Invent destinations (that has been done on numerous occasions) which you have no intention of visiting. I am aware of one prominent figure who regularly claims substantial grants on the basis of liberal overseas travel, but who is a classical xenophobe and spends what spare time he has in a caravan on the Norfolk Broads, and sometimes with them too. His great excuse is that frontier officials at passport control rarely bother to stamp a passport in Europe these days, so his invisible travels are not identified by an empty passport. He keeps it in his trousers' back pocket during the day, so that it gets to be well-worn and grubby-looking, and props it behind the clock in his drawing-room when he entertains. On the wall are colour pictures of favourite cities, at least one of them misidentified, which he claims to have taken himself but are actually copies photographed out of travel brochures and part-works on *Exploring Today*.

An example of custom-built apparatus which was published in the letters column of *New Scientist* magazine. It was built to investigate 'the nature of randomness' – from which one's own conclusions may be drawn.

By the end of the financial year all the grant has to go. There are plenty of ploys that can be used here, of course. Back-dated invoices can allow the supply of goods to be fitted into one year, even if they do not strictly belong there. Some manufacturers of equipment will provide a cash discount in the hand in exchange for the placing of an order that might otherwise have gone elsewhere, and apparatus is always so grossly over-priced that there is plenty of lee-way for an enterprising negotiator to line his back pocket under such circumstances.

Often at the end of the financial year, you will find anxious girls standing in line at an enquiry desk ordering electronic word-processors, letter-openers, elaborate electronic calculators ... anything to use up the cash. This apparatus is usually stored away gathering dust, and it amounts to over £4,000m in Britain alone, according to one reliable estimate.

The hyperbole question is an interesting one. Sometimes people give the impression that they are only after the money if it is absolutely convenient for the funding agency. Their explanation reeks of altruism and modesty. No Expert should ever be tempted to indulge in that kind of exercise. The grant must be seen as vital for an imminent breakthrough, and there should be plenty of allusions to world-wide repercussions if the grant is forthcoming. A few references to implications for cancer research, feeding the Third World, reliable contraception, recycling or Appropriate Technology can work wonders. On the other hand, objective summaries of what you have in

> If a project costs too much, you have a "suboptimal cost profile" on your hands as well as a "cost overrun." If a market researcher goofs, he hides behind a "demographic skew." No longer do we have depressions or even "recessions." The latter is now a "recedence," a "retardation," a "pause," or a "rolling adjustment." Inflation is "price adjustment."

mind are a fatal mistake. In the United States there are professional organizations which undertake to write applications to a strict formula, and (as long as you are sure the formula is not going to recur in someone else's application statement) this is a clear advantage, and allows you to dispense with any form of integrity whatever.

The fact that grants should be progressively larger

each time they are due for renewal is not, as you might imagine, because of inflation. Grants must actually increase, in real terms. A good estimate is 15-20% over inflation, though of course it is perfectly possible to make that 150-200% if the situation demands it. The admirable C. Northcote Parkinson has shown with unblemished accuracy how it is easier to apply for a large grant than for a small one, and the implications of Parkinson's *Law* should be taken to heart by everyone out for funding.

Finally, how should you describe your apparatus? No-one argues about an unambiguous term like a calculator or a ball-point pen. But some items can be up-graded. Thus, instead of an electric typewriter it is better to put in for an *electronic micro-chip word processor*; rather than a camera make it an *image-generating reprographic optical documentation facility*. You can then order the cheaper version, collect a faked invoice for it, and the friendly rep. will present you with the difference (less commission) for your own use.

> Dutchman Henk Bredeling uses a Commission car and chauffeur every day to travel from Brussels to his home in Holland.
> During 1978 Mr Haferkamp spent £130 one night for an hotel bed.
> One man accumulated an advance on expenses of £1,600.
> Another Commissioner is alleged to have spent close on £400 a week on flowers for his office.

It is perfectly possible to use up a grant in expenditure for pleasure, rather than for work. Take the laboratory whose technical staff want a microwave oven to warm their meat pies at lunchtime and make soup on cold mornings. The obvious advantage of buying one with a grant is that it does not cost anything. But there is a second benefit: a microwave oven ordered for a laboratory will be bigger than a normal one, and probably somewhat better too. It will also cost five times as much as the basic article, but that is normal in the supply of laboratory apparatus.

127

However, it is not very likely that approval would be given for a form which said:

RE SUPPLY ONE MICROWAVE OVEN
PURPOSE PIE-WARMING (LUNCHTIME)
AND SOUP PRODUCTION
(CHILLY MORNINGS)

Instead, careful wording will make the device look abstruse, vital and somehow too mysterious to be worth investigating:

REQUIRED – ONE ULTRA-SHORTWAVE
THERMAL GENERATION CABINET
FOR FOOD TECHNOLOGY
EXPERIMENTS AND STERILIZATION
RATIONALE DEVELOPMENT

and through it goes. The cost is $4,500 instead of $850, but who cares? It is only grant money.

Another research group want a snooker and pool table for their downstairs laboratory, to while away the many unoccupied hours that are the hall-mark of modern high-power investigations. Clearly they could always order one themselves and foot the bill, but such things are costly. With a little thought the item becomes a SLATE-BED INERTIAL STABILITY BENCH. And if you wanted a few sets of balls to go with it, then they could easily be entered as NON-DIGITAL HORIZON-DETECTING ORIENTATION INDICATORS since they will obviously run slowly down-hill if the table itself is out of true.

The best example that has been drawn to my attention was the group who needed a replacement piano for their social club-house. Last time I heard, they were toying with AUDIOFREQUENCY HARMONIC TONE GENERATION FACILITY, which stands a good chance of success, I'd say.

And meanwhile you can always order the sugar for morning coffee – SUCROSE, ANALAR GRADE – and top up the tank of your car with a few carboys of PETROLEUM SPIRIT, RECTIFIED. At home you can run for years on liberal kitchen tissues ordered as LABORATORY CELLULOSE WIPES, and washing-detergent which began life as SURFACTANT BIODEGRADABLE DETRITUS EMULSIFIER. Truly, there is no limit. The laboratory world is full of controlled-temperature rooms curing illicit casts of famous sculptures, and large incubators turning out duty-free wine and beer by the tank-full; half the holiday snaps are taken with the institute's cameras and processed using their equipment; and the loan of microbuses (which quickly become cara-vanettes for family jaunts) and video-recorders (to make a programme about your vacation) is enough to save everyone a few grand each year.

It is as well to remember that once a grant has been allocated there is no come-back. The expenditure itself is not investigated. The only difficult part is getting it to begin with, and it helps if staff under-stand how important it is to create the right impres-sion. Here is one genuine example:

```
TO ALL MEMBERS OF THE DEPARTMENT
VISIT BY THE UNIVERSITY GRANTS COMMITTEE
TECHNOLOGY SUB-COMMITTEE
THURSDAY, 11th FEBRUARY, 1971

    The above Committee will be visiting the Department
on Thursday, 11th February between 11.15 a.m. and 12.30 p.m.
and between 2.00 p.m. and 3.00 p.m.

    We clearly want to give a good impression and for
this reason all members of the Department are asked to
implement the following:

1. All laboratories are to be cleared up and rubbish or
equipment not used is to be stored away without, however,
giving the impression that any special steps have been
taken to tidy up the place.

2. All research workers are asked to work in the
laboratories during the period of these visits simulating
'feverish activity'.

    The intention is that we should give the impression
of being a highly active Department which looks after its
resources and utilises them to the utmost.

                            S.A. TOBIAS
```

The Birmingham Students' Newspaper, in report-ing the above memorandum, published a suggestion that it betrayed a very 'peculiar' attitude. It may look that way to the naive student. But to the initiate who understands how the system works, it is exactly what you would expect.

In certain respects the more outrageous the project, the better its chance of success, since there will be fewer people qualified to spot the loopholes. A congressman from Arizona recently documented some current projects, and they will be a fitting tribute to the art of grantsmanship as well as an encouragement to those who are following on:

+ $20,324 to study mating calls of Central American toads.

+ $19,300 to find out why so many children fall off tricycles.

+ $121,000 to discover why Americans say *ain't* instead of *isn't*.

+ $375,000 to see if frisbees might have other uses, such as carrying flares at night.

+ $20,000 to investigate the blood groups of Zlotnika pigs.

The range of possibilities is unlimited, and it is time that a special prize fund was instituted to give an extra grant as an incentive for the wildest and most improbable scheme to be heavily funded in the coming years. But the competition would have to be framed in carefully-expressed rules. One recent innovation, for example, was for a sundial that could tell the time even in the dark. It had an electric lamp fitted to it. As it happens the designer intended this as a joke, but it has been taken so seriously by most other people that it is now due to go into production. A contestant for the proposed 'zaniest grant in the world' competition would have to be genuine, therefore, and not a joke.

The only problem is that it is becoming increasingly difficult to tell them apart.

CHAPTER EIGHT

Innovation from Without

Why do we need elaborate and expensive research institutes? The answer is simple: without them there would be no high-technology research, no modern era, no science-based civilization which is poised on the brink of establishing a new age of complete knowledge. So say the Experts.

The truth is that those elaborate centres are needed only by Experts who wish to perpetuate themselves, their status and the ephemeral pronouncements they produce. Although it is undeniably popular to imagine that all the great discoveries have been made in a monster ultra-modern centre, that view is erroneous. Perhaps 85% of all establishments are there in the rôle of 'mutual self-perpetuation'.

The most telling way of demonstrating this is to look closely at the process of innovation. Most of the great technological developments on which our modern society depends (from colour photography to photocopying, from aircraft to the long-playing record) have been developed by part-timers, or by people who were working in the traditional manner, for themselves and not for any large organization. And many of the greatest discoveries of all – such as penicillin and plastics – were made by accident anyway.

The part played by the large research organization is very small by comparison. Most of them are involved in playing the game, aiming for international prestige and materialistic might, setting up codes of conduct which they then build laboratory blocks to implement.

Let us look back over the recent history of technology, where we can find classical examples of individual research which has (in practical terms) put the organizational giant somewhat to shame. Let us take

131

three categories of research: (a) where the importance of the discovery is great, but the investment was low; (b) where a scheme was developed and brought to fruition even though informed opinion was set against it; (c) where individuals in unrelated specialities – from violin-player to office worker – made major break-throughs which altered our lives for the better.

a) Discoveries on a shoe-string

The idea for transistors was conceived before the Second World War when the head of research at Bell Telephone Laboratories launched a study of solid state physics, concentrating on semi-conductors (the materials of which transistors were made). William Shockley and S. O. Morgan headed the research to begin with. The war prevented the carrying out of many practical tests, which would have proved whether Shockley's theories were right or wrong. But after the war the first experiments were a disappointment. The amplifiers did not function as Shockley had theoretically predicted. To many people it looked as though transistors were doomed before they started.

And it was then that the low-budget individuals came in. One was John Bardeen, who worked out a new approach to the theory of transistors; and the other was Walter Brattain, who carried out some experiments with electrical fields applied to semi-conductors from the outside. Eventually the two men worked together, and managed to show that you could modify a current flowing *through* a 'transistor' model by means of a current applied to it from the outside. That gave rise to the transistor revolution, and from that came the ubiquitous (if grossly over-rated) micro-chip.

Throughout all this vitally important work they used very simple apparatus. Their most complex item was an oscilloscope. With less than £100-worth of apparatus the world was changed.

Television is one invention which people know to be the result of a single individual's work – John Logie Baird, the Scot. Well, that may be what they think, but they are wrong. Baird had that gift for catching the corporate eye of a community, and it is certainly true that after 1929 there were crude TV transmissions using his system. But it was always obvious that a mechanical TV was bound to have serious limitations, and Philo Farnsworth (a self-taught American based in Los Angeles and San Francisco) demonstrated a complete electronic television system as early as 1927, when he filed his first patent. His image dissector tube was a vital development in the history of TV and although there were other workers making parallel discoveries (of whom Vladimir Zworykin was surely the most important) Farnsworth's energy and independence did much to give us TV as we know it. I am not suggesting that our habit of squatting in front of a television set, staring into the tube like hypnotized rabbits night after night, is a laudable thing. I doubt whether Farnsworth imagined that we would. For him the purpose was to take useful information across distances, which is a very different approach to filling up air-time with whatever happens to be cheap and available.

A parallel development to TV is, of course, radio. The first person to transmit speech was Reginald Fessenden who achieved this in 1900, working for the American Weather Bureau. Another independent worker, Lee de Forest, invented the triode valve (he may have pinched the idea from a lecture Sir Ambrose Fleming gave to the Royal Society in 1905). Edwin Armstrong, who discovered that triodes could be used as amplifiers, was a college student at the time, and then of course came the independent spirit of Marconi himself.

Radio leads us fittingly to records. Tape-recording was invented by Valdemar Poulsen, a Danish telephone engineer who made his own breakthrough by working in the evenings and at weekends, purely as a hobby. He patented the device as long ago as 1898, and in the 1930s the idea was further improved by the research work of the gifted Marvin Camras, who perfected a greatly improved wire recorder. As you

133

may anticipate by now, he was a student (at the Illinois Institute of Technology) and his research was a hobby too.

But what of big physics? Here too the unfettered individual has led the way. The cyclotron, that embodiment of high-technology, was not invented in a huge laboratory with innumerable willing staff on hand. In the words of a report in *Scientific American:*

There was an ordinary wooden kitchen chair on top of a physics laboratory table ... on either side of this chair stood a clothes tree, with wire hanging on the hooks which would normally hold hats and coats. Between the two poles the wire was suspended in loose loops. The loops went all round the chair, on the seat of which was an object about the size and shape of a freshly baked pie. It was made of window glass, sealing wax and brass.

What about computers? It is hard to imagine any development more unassailably complicated-looking, more important or more high-technology oriented than computers. They had their roots in the mechanical Difference Engine and the Analytical Engine designed (but never built) by the eccentric Charles Babbage in the nineteenth century. The idea of punched cards derived from the Jacquard weaving workshops in Lyons, and was taken up by an American, Herman Hollerith, in the closing years of the nineteenth century (he used them for the 1890 United States Census, for instance). Hollerith formed a company which ended up as IBM, but his important early work was all done as an individual. A step towards an electronic computer was taken by Friedrich Züse, when he was still a student in Berlin. Unfortunately he then tried to gain support from the German authorities, who did help him a little, but as soon as he proposed the development of an electronic computer the Experts cancelled all support. The modern computer later arose from the insight and industry of these individuals.

Computer technology takes us to the realm of automatic photo type-setting, which is revolutionizing the printed word. Watching a photo-typesetter in operation is an astonishing experience. The idea was

134

first mooted by a Swiss engineer, C. E. Scheffer, who adapted traditional hot metal machines into a primitive form of photosetting device. But the first wholly new photo-typesetter was developed (in their spare time) by two telephone engineers, René Higonnet and Louis Moyroud, who built the apparatus in their homes. A cathode-ray tube was brought in to a more refined version of the principle, invented in England by R. Mckintosh and P. Purdy, respectively a journalist and a photographer.

Then of course there is the hovercraft, brainchild of Christopher Cockerell, an electronics engineer turned boat-builder. He made his prototype test rig from two coffee tins and a hair-dryer poised on a set of kitchen scales. You might next turn to rockets, conceived by K. E. Ziolkovsky whilst a schoolmaster, developed by Hermann Oberth in his book *The Rocket into Inter-Planetary Space* (which was published whilst he was a student, in 1923), and first tried out successfully by the pioneer of liquid-fuelled rockets, R. H. Goddard, in his spare time. Even the German developments of the Second World War were the result of studies made by an amateur group of enthusiasts, the VfR, Society for Space Travel (of whom von Braun was a leading member).

If rockets are somewhat inaccessible objects, and hardly part of your day-to-day life, then we can bring xerography – dry photocopying – onto the scene. This important idea was conceived by a patent lawyer, Chester Carlson, who was bemused by the difficulty of obtaining copies of official and legal documents. His first prototypes were made at home, in the basement.*

b) Working against the stream
The plastics industry is an important part of modern life. It began with the world's first thermosetting

* Blakeslee, H., Atomic Slingshot, *Science Digest,* April 1949.
Schuler, L., Maestro of the Atom, *Scientific American,* August 1940.
Jewkes, J., Sawers, D., & Stillerman, R., *The Sources of Invention,* Macmillan, London, 1969 edn.

135

resin, which became known as bakelite. All the earlier research work, before this plastic was discovered, suggested that you had to control the temperature of the mix and keep it well below 100°C. But Leo Bakeland, the man who invented 'Velox' photographic paper (itself a revolutionary product), began to suspect that this idea was false. In a barn at the bottom of his garden in Yonkers, he set out to challenge traditional thinking. Not only did he raise the temperature of the reaction, but he discarded the normal idea of using acids as the 'condensing agent' in the process, preferring to try alkalis instead.

His ruse worked. He received patents in 1909, set up his own company as the Bakelite Corporation in 1910, and the rest (as they like to say in all the best television documentaries, so I am told) you see around you to this day.

Plastics come from oil, and the catalytic cracking of crude petroleum lies at the basis of that industry. Finding a material that encourages petroleum molecules to break down to form the most valuable fractions – the light oils and petrol itself – was not difficult. The problem was to find a catalyst that was not quickly poisoned. It was a subject that caught the attention of a young racing driver, Eugène Houdry. In the 1920s he investigated many alternatives, and was told on many occasions – to quote his words – 'the research men were absolutely certain that catalysts could not be regenerated.'

He considered over 1000 possible materials, and was eventually convinced that the answer lay in activated clay, perhaps because it was already in use as a bleaching agent for lubricants, and it had been said by the suppliers that it could not, in any circumstances, be reactivated. In the end he made it work very well indeed and so, as an individual, Houdry founded single-handed the modern era of oil refineries.

In the medical field, the discovery of insulin by F. G. (later Sir Frederick) Banting and C. H. Best probably owed something to the fact that Banting was told repeatedly that the idea was bound to fail.

136

Jet engines, in much the same way, developed from the work of Frank (later Sir Frank) Whittle when the Experts – with some irritation – informed him that he did not know what he was talking about when it came to turbine design.

No-one knows how important it is for an innovator to be told by an Expert that his ideas are pointless and that his proposals will not work in practice. But I am inclined to believe that this is exactly the stimulus that many people need to encourage them to give of their best.

c) The Innovator from outside

Great discoveries that have been made in conditions of spartan frugality have given our cushioned, easy life many of its pleasantest features; and researchers who go against the stream of Expert opinion have always appealed to us for their bravery and cheek.

Take Kodachrome. As a system of colour photography the Kodachrome principle represents a valuable development, and I do not doubt that you can easily picture those research workers industriously striving towards their goal, all in serried rows of uniformed employees wearing red-and-yellow identification badges and clocking in on time. How far that is from the truth!

Rudolph Fischer with his co-workers laid down the theory of multi-layered colour films before the First World War. The difficulty was that the ideas (which were so simple in theory) proved to be impossibly complex to realize in practice. Fischer's Berlin patent of 1912 specifies the three-layer monopack emulsion which is used by all the major film manufacturers today, but when attempts were made to produce it, the colours ran from one layer to the next and the result was highly unsatisfactory.

All the while, however, there were two schoolboy friends in New York State who liked to carry out photographic experiments in their spare time. They made some accessories and modifications that would improve projectors, for instance. Their names were Leo Godowsky Jr. and Leopold Mannes, and they

137

were inseparable throughout their schooldays. When they left High School they both decided to study music, and this they did; Godowsky went to the University of California, whilst Mannes left for Harvard.

Whenever they were home for holidays, and throughout the long summer vacation, they would experiment together. For a time they earned a living by teaching music. In the evenings and at weekends, they would experiment in Leopold Mannes's kitchen. Eventually they succeeded in producing their first film in which the colours were all present in a single layer. It was 1923.

The two musicians eventually became concert performers, and used to tour Europe playing to packed houses who little suspected that in their luggage they carried a reasonably complete set of chemical laboratory apparatus and a full range of reagents, so that they could carry on experimenting in the hotel rooms each day. They soon succeeded in finding the right way to attain full colour saturation, whilst preventing the colours from running between one layer and the next.

So two musicians – professional, concert performers at that – perfected modern colour photographs. The work of these two violinists has affected the lives of everyone in one way or another.

The ball-point pen is a symbol of modern convenience, too; in Third World countries (which Experts refer to, in the new fashion, as 'LDCs' – Less Developed Countries) there is no more universal token of tourist bartering than ball-point pens brought in from the ODCs.

Synonymous for 'ball-point' is 'biro', of course. And few will be surprised to learn that the inventor of the device was Ladislao J. Biro, a Hungarian, assisted by his brother Georg. They carried out their original work in the back room, making their own machinery to produce the fine components they needed. By 1938 they were applying for their first patents, but when the Second World War broke out they emigrated to Argentina. From there the ball-point became a world success, partly through the enterprise of an American named Milton Reynolds, who managed to redesign a biro pen he had bought in Buenos Aires in 1945 just enough to get round the patent protection and put it on the market before the Biros themselves could ship their products into the United States.

Ladislao Biro saw his surname enter the English language. And his profession? He was a writer, and an artist (working both in sculpture and paint).

A mainspring of modern navigational technique is the gyrocompass. Many people had foreseen how useful such equipment would be, but it was once again a problem of converting theoretical understanding into the crisp reality of a device you could manufacture and use. The answer came through the work of an art historian named Anschütz-Kaempfe. He was left a legacy by the death of his adoptive father, and decided it would be fun to travel to the North Pole by submarine. There was one important practical difficulty, however; and that was navigation. He experimented with a couple of gyroscope compasses, but without much success as he realized that they wandered too much. So he designed an alternative version in which the gryroscope pointed permanently to the north, as a real substitute for a traditional magnetic compass which was, however, unaffected by the vagaries of the earth's magnetic field. Near the north magnetic pole, the lines of force plunge with little sympathy for navigators into the earth's crust, but Anschütz-Kaempfe's gyrocompass would, he knew, be immune to that tendency.

Gyrocompasses are now world-wide and are relied on in a host of navigational applications. The fact that they were first successfully demonstrated by an art historian is somehow intriguing. On the other hand, it is also just what one might have expected.

And what of the safety razor? King Gillette, who invented it, was convinced that there was an alternative to the traditional open razor, and that a thin, disposable blade was the core of the solution. As he later said, 'If I had been technically trained I would have quit, or probably would never have begun.' As it happens, he did not suffer a burden of specialist training to detain him at the bottom of the ladder, and so set off to produce one of the handiest domestic advances technology has given mankind.

The idea arose 'in a flash', as was later said, when he was shaving one morning in 1895. He rushed to the store, to buy a small vise, some brass components and a length of spring steel from a clock mainspring, and with them he fashioned the world's first safety razor. The patent was issued just nine years later. For years Gillette was unable to find any support, either in spirit or financially: the metallurgists were entirely convinced that steel edges could not be tempered or hardened as Gillette desired, and his friends thought he was a little crazy to bother to try. Eventually his persistence won through.

And Gillette's profession? He was a travelling salesman in bottle-stoppers.

If you were invited to nominate a single instrument by which modern science was identified, a symbol of the research laboratory, it would surely be the microscope. This is no place to attempt to recount the tortuous story behind the microscope's development, but we can of course refer to the father of microscopy, the great Dutch pioneer, Antony van Leeuwenhoek. Single-handed, he discovered bacteria, documented the protozoa, made some seminal (*sic*) discoveries on reproduction which included studies of spermatozoa from a host of animals. He was the first to document the nucleus, and in his own workshop he not only perfected the highest-magnification lenses ever made at the time, and made the little microscopes in which they fitted, but he even refined the metals he used from the crude ore itself. His work was prodigious, and it is without question in the forefront of the realm of microbiology. Van Leeuwenhoek was visited by royalty and by heads of state from many nations. Yet he was no educated man: he earned a living as a draper's salesman and town official, and did all his scientific work as a hobby throughout his life.

Even William Herschel, the great astronomer, was

in real life a musician and composer. And it was students who invented domestic gas-refrigeration on one hand and polarizing filters on the other. It is recorded that every type of automatic gun has been the work of civilians, individual inventors in each case; and the automatic telephone dialling system which revolutionized communications was the work of an undertaker. Among the leading pioneers of the rayon industry were a bank clerk and a glass blower, whilst cellophane was perfected by a dyer.

Other developments from the zip fastener to electron microscopes, from titanium to the Wankel engine, from power steering to air conditioning, and from the hypodermic syringe to Cinerama, chrome-plating and the cotton-picker ... they have all been the work of individuals rather than corporations. Even in aerospace the solitary approach has been the mainstay. Virtually every important development of aviation comes into this category.

Miller and Swers, in their study of the history of the modern aircraft, have argued that only two inventions (each one of a kind of flap for aircraft wings) have resulted from the employees of the aircraft manufacturers' research establishments. Many of the others were the work of independent individuals.

So convinced have we become of the might of the great institution, and the idea that massed hordes of specialists will produce the goods, that we have ignored the facts.

Small is Abhorrent

All Experts know how impressed people are by materialism. They know how, if a salesman calls to your door on his push bike with his trousers tucked into his socks, he will receive a very lukewarm welcome compared to how he would automatically be treated if he arrived in a large saloon or a low-slung sports car and wearing a sleek suit. So staff in the institutes are provided with the most prestigious microscopes available (most prestigious, mind, not necessarily most *efficient*); they use masses of costly disposable materials

141

from petri-dishes and scalpels to gloves and gowns. They have rooms filled with stores that most visitors would find overwhelming. They are giddy with golf-ball typewriters where a simple note-pad would suffice; lost amongst calculators, video machines, colour cameras and electronic bleepers to summon them when the tea is ready.

It is accepted nowadays that the manufactuer's name-plate on an instrument can be directly related to the status of the institute itself. It is a matter of pride to be taken round a research laboratory and be told: 'This device costs twice as much as a Rolls,' even if there is a thin layer of dust on the controls and nobody in the building knows how to use it anyway. There are some laboratories where you can find electronic balances, lasers and centrifuges literally piled on top of each other for want of storage space. Few people ever use them, indeed it is common for most of the staff to know nothing about the purpose of a given instrument – and even if they did, it is a cardinal principle of all such devices that they do not replace anyone on the staff. So in personal, convenience terms they are irrelevant, and in the most hard-headed economic terms they are not in the least cost-effective. But they do boost the institute's corporate image, which counts for more than anything these competitive days.

The one question you should never ask is: 'Exactly what results have you come up with?' Results are printed out on long folded computer sheets covered with numbers and symbols that neither you, nor they, can possibly decipher. There are an awful lot of results. But what you *meant* by your question was: what recognizable benefits have been derived from it all? The enquiry is never perceived in that way, however, since the answer could then be a cold 'nothing much', which is more than any Expert can bring himself to admit.

There is much lordly talk of 'world-beating' this or 'revolutionary' that, but when you get down to bare facts and look at the conceptual innovation or the practical breakthrough there is simply nothing there.

142

On several occasions I have spent time with departments like this, searching for just one item you could celebrate, and found nothing. There are plenty of results – repetitive calculations pinned to the wall and the usual list of publications in over-priced journals. But no actual output that matters.

One example was the video programme that had been put together to extol the virtues of the department to all outsiders. I had heard of it, but never actually managed to see it. Nor had anyone else. It took a good deal of persuasion to have the tape dug out and dusted off. 'This is a copy that we keep to show people who really want to see what we do,' I was told. 'The original is kept safely. Too precious to show.'

But the entire production was without any question the worst film of any kind I have seen, with the possible exception of a holiday movie I had once been sent by an enthusiast who taped an 8mm camera to the front of his car and drove around Scotland at some speed, the blurred tartan images whisking past the Mercedes-Benz emblem on his radiator that otherwise filled the screen for all of fifty-five minutes. The academic offering I was shown was badly lit, unedited, with sepulchral voices that echoed around the room without joining into sentences at all; and the scenes of experiments were disjointed and so badly set up that all you could see was a navy-blue screen with the occasional darker blob making indeterminate movements.

The comparison I have just drawn gives the game away, of course. These productions *are* holiday movies. The camera (which is naturally the most expensive closed-circuit TV camera you can buy) is waved around helplessly, wide-eyed and manic faces jabber into its lens, and complex apparatus designed for taking experimental shots is malfunctioning because nobody can understand the instruction book.

The point to emphasise is that if you were to make such a film yourself, using a home movie camera worth a few pounds, it would be condemned as amateurish and disjointed. When an Expert does it,

143

however, that does not happen:

FIRST because he is an Expert, so no-one is going to argue;

SECONDLY, since if you cannot quite make out what is going on, well, that shows how out of touch you are.

It is a salutary revelation that the worst video films in the world are those made in laboratories, and it is an interesting thought that in every case the most expensive and up-to-date apparatus was used in making them.

People have been convinced that it is only the well-heeled and impressive-looking foundation that can carry on high-powered and productive research. This seems logical enough when Experts present it with the words: *'Obviously* we need all that backing if we are to remain in the forefront of knowledge.' No-one tumbles to the fact that 'forefront of knowledge' and 'money' do not really have anything in common at all.

This is just as well for the cult of the Expert. If the public were to realize that almost all the major discoveries were made outside the well-funded institute, then we might all be a mite more sceptical about putting so much faith in capital-intensive research. That great industry, petrochemicals, was analysed some years ago and in the words of John Enos:*

The most novel ideas – cracking by the application of heat and pressure, continuous processing, fractionation, catalysis, the generation of catalysts, moving or fluidized beds – occurred to individual inventors, men like Dewar and Redwood, Ellis, Adams, Houdry and Odell, who occasionally contributed their talents but never their employment or loyalties, to the oil companies.

Now, if the principle is true, you would expect that the greater the development, the less it would cost. Let us select the greatest scientific idea of all in recent times. What would it be? Penicillin, perhaps? That certainly qualifies. The discovery was made by

* Enos, J. L., *Petroleum Profits and Progress,* M. I. T. Press (1962).

144

Alexander Fleming using a single dish of agar on which a spore of *Penicillium notatum* happened to settle one day in 1928. Even though he originally thought the mould was something else (it was wrongly thought to be *Penicillium rubrum* at first, and it was only after a reappraisal that it turned out to be *P. notatum,* a fungus that had been discovered many years before by Westling in Scandinavia, where it was growing on a heap of decaying hyssop) and then gave the name 'penicillin' to the growth medium in which the fungus grew, rather than the antibiotic it contained (which was a clear error), and certainly did not invent the word 'antibiotic' (which had been coined earlier by Maury in a wholly different context back in 1860), he nevertheless proved how powerful penicillin was and showed that it would control a range of bacteria. And how much did it cost? That is hard to estimate, but since he worked with ordinary flasks and petri dishes I imagine that it must have consumed materials costing around £10 (i.e. less than $25 altogether).

Penicillin was maybe the greatest discovery, but the greatest idea of twentieth-century science must surely be the theory of relativity. Almost everyone in all walks of life has heard of it, and it produced a revolution in science of unparalleled magnitude. The effects of relativity live with us today, and the whole of thinking in physics was set into a new light by this breathtaking hypothesis.

Now, if the inverse relationship between cost and inventiveness applies (and according to the thesis I have put forward that ought to be the case), then it would be sensible to assume that a truly momentous discovery like that should cost less even than the discovery of penicillin.

This is indeed the case, for Albert Einstein was working as Assistant Clerk in the Swiss Patent Office during the years 1902-1909, and when his paper on the special theory of relativity appeared in 1905 it was the result of his own spare-time work. Thus, the greatest scientific theory ever cost not much more than the price of a stamp.

CHAPTER NINE

Communicate or Die

The incisive mind will know full well that Experts do not communicate anything. They confuse, they perplex, sometimes they confound; but they do not actually communicate. Yet their pronouncements have the full-blooded vigour and complexity of a Court judgement played at the wrong speed. So, although it is true that there is nothing of substance in what Experts say, they do send out plenty of data in their own way.

There are several essential rules which underlie the Expert's peculiar brand of quasi-communication:

a) Never use one word where many will do
It is a profound error to aim for simplicity of language and ease of style. Any Expert will tell you that. A British Medical Association report examined clarity of exposition, and gave a most meaningfully viable summary of the ongoing scenario:

Clarity of exposition depends on clarity of the pictorial and of the verbal components as well as the arrangements of the contents which is concerned with level of difficulty in relation to the existing understanding of the learner.

Experts should try not to say *find out* when they can manage with *cause an in-depth investigation to be shortly made with an ultimate view to ascertaining the real-time parameters governing the situation*, nor should they make do with the terse *until* when they can wangle *towards such due time as*.

THE GOLDEN RULE IS SUMMED UP IN THE WORDS
A plain man calls a spade a *spade*. The true Expert calls it a *vertically momentum-actuated pedistical spatulate alluvium disinterment facility*.

b) Never use a short word where there is a perfectly useful lengthy alternative

Never use *although* when there is *notwithstanding*. Do not be trapped into saying *red-cell* when you can say *erythrocyte*, and avoid *thought* as long as you have *conceptualization* in your mind. One of my personal favourites is the correct term for *tooth-grinding*, which is *temperomandibular-joint syndrome*. But take care, for there are pitfalls for the incautious. Though it is an instinct for any Expert to substitute *anticipate* for *expect*, it is unlikely (as Sir Alan Herbert once said) that anyone would interpret *John and Jane anticipated marriage* to mean the same as *John and Jane expected to be married*.

c) *Invent new words whenever postulacious*

I have already done this, in the true spirit of my subject in this book, with *nonscience* and *fashionism*. The trouble with those words, though, is that they can be easily understood. It is far better to head into abstruse realms, and derive some new nugget like this:

Toputness is system environmentness ... and storeputness is a system with inputness that is not fromputness. Disconnectionness is not either complete connectionness or strongness or unilateralness or weakness, and some components are not connected ... in other words, feedinnness is the shared information between toputness and inputness where the toputness is prior to the inputness.

That is from an American research report entitled *Educational Theory Models*. There is ample scope for newcomers, still: why not a *microorganismic pathogeneticism* instead of a *germ*, or a *intrafactorialism* instead of a *number*? Those are just a start.

d) *Keep the grammar complicated*
As a management training summary once eloquently said:

The cognitive continuum is concerned with objectives

147

related to knowledge and the intellectual abilities and skills, rising from comprehension to evaluation. The affective continuum covers the range of behavioural responses, from passive acceptance of stimuli to the organization of taught values into a complex system which constitutes the whole characterization of an individual.

In *relationship to the employment of personnel,* another writer has acknowledged that:

In the second place there are grounds for thinking that the availability of analytical assessments of jobs would facilitate the preparation of grade-descriptions for a new structure in a situation in which the allocation of jobs to grades at the stages of implementing and maintaining that structure would be undertaken by whole-job procedures.

And the need for the public to be brought into the picture, especially when there are a lot of new data to put into practice, has been poetically conceded:

Diffusibility of knowledge throughout the environment in which the families are to move is essential if the full expression of their potentiality is to become explicity in action. Facts pertaining to experience of every sort that the family is in course of digesting give the context and the full flavour of consciousness to their experience.

The establishment of minimum constraints for an optimization of free growth combining elements of user-design for both the individual and the community, forms the basis of this project. The basic order devised is intended to establish a democratic inter-change between human and technological factors. The order devised will stimulate

---

Reading ease (RE) of letters published in *Nature* over half a century

Year	No. of letters	RE (s.e.)
1930	8	32.1 (6.3)
1940	10	29.1 (2.6)
1950	29	16.5 (2.4)
1960	76	18.0 (1.5)
1970	59	15.0 (1.6)
1980	41	12.8 (2.1)

Are scientists expressing themselves even less intelligibly? Findings published in *Nature* in 1981.

148

multiplicity, multiformity, micro and macro relations; all expressed through logically derived dimensional and functional modules themselves articulated by a system of guiding lines.

Sir Ernest Gowers' *The Complete Plain Words,* revised in 1973 by Sir Bruce Frazer, quoted the above paragraphs and admitted a complete inability to understand what they meant. Really it is quite simple. The first paragraph means that datum-translocation interface mechanisms in the family expressive stricture set *vis-à-vis* community oriented experiential input on the basis of received cerebrization require pragmatic reassessment in an on-site real time situation. With that as a guide, I am sure I can leave you to work out the second paragraph for yourselves.

e) Cultivate the craft of vagueness
You must be careful here. Many people have tried to imitate the true Expert's verbiage and have trapped themselves instead.

There was the lady researcher who was trying to write about a vitamin named *B*-carotene. She spelt it out in a private way she derived for herself, so that it became *Beta carotene* instead. Thus, instead of writing about a vitamin she made it look like a species of turnip. Another commentator was attempting to throw in a few medical terms in a paper about Africa, and came up with *elephantitis.* I expect he meant *elephantiasis*, which is a condition of enlargement of parts of the body caused by a parasite in the lymph glands. By writing about elephantitis, he was describing something quite different: it means, as far as it means anything, that your elephant is inflamed.

Beetles which seemed to be moving normally were counted as alive, beetles moving but unable to walk as moribund, and those not showing movement within three days as dead.

As it is important not to muddle a *micrometer* with a *micrometre*, it is just as necessary to keep *penicillin* distinct from *Penicillium* or a *penicillus*. Do not confuse *Pediculus* (an insect) with *pediculosis* (a disease). Remember that the simple word 'sterile' has half-a-dozen entirely different meanings. Try not to be confused that a term like 'osteogenic' is nicely ambiguous too, it can either mean *bone-producing* or else it can signify *produced-by-bone*. There are such delights as 'anaemia', which means literally *without any blood*, and 'comatose' (much beloved by sectors of the medical profession) which means anything from *very much alive* to *very much dead*, depending on who has their finger on the switch of your life-support machine, and how much they need your innards for a project.

A subject worthy of special study is where Greek and Latin collide with sense. As an example of what I mean, consider the prefixes meaning 'above' and 'below' – in Greek respectively *hyper* and *hypo*, in Latin *ultra* and *infra*. There are strict categories where they are used; sometimes they are synonymous and sometimes mutually exclusive.

For instance you can say 'ultrastructure' and 'infrastructure' which both mean exactly the same thing. In the same way, you can utilize 'ultrasensitive' or 'hypersensitive' without giving rise to the merest flicker of an inquiring eyebrow. With some other words you only ever use the Latin or the Greek prefix, and the use of its opposite number would look like base illiteracy of a very odd kind. Thus you have 'ultraviolet' and could never, without risking your reputation, write that as 'hyperviolet'. On the other hand, and for no particular reason, we always say 'hypothermia' and would never use 'infrathermia'.

Are you tempted to rationalize all that and say that the Greek prefix goes with the Greek root … ? That won't do. Look at the ambiguous alternatives which began the previous paragraph. And what do you make of the case of 'ultrasonic' and 'hypersonic'. They are both admissible words, yet they have distinct meanings. Ultrasonic means 'at a frequency above that which is audible', whilst hypersonic means 'travelling at a velocity greater than that of sound waves'. This last word can be replaced by its synonym 'supersonic', of course, which opens up a range of new alternatives from superviolet and superthermia to a whole new range of meanings for superstructure. It can only be a matter of time before we see stories about *Hyperman in his Inframarine Hypomobile.*

f) Master the pronounciation problem
What *is* important here is to follow the fashion at your place of – well, not *work* exactly, place of employment. The difficulty lies in the range of possibilities. For example, take the word *carcinogen.* It looks simple enough, but there are several alternatives that are current just now and it is important to stick with whichever one happens to be used by the boss. The main versions are:

a) car'cinogen – where the second syllable is almost lost, and the word sounds like 'car-snojen';

b) carcî'nogen – in which the first syllables sound like 'car sign';

c) carcin'ogen – which is as (b) but the second syllable is sounded to rhyme with 'sin' instead of 'sign'.

At one time it was possible to assign geographical distributions to these (option (b) was used a lot in London, for example; (a) in the English provinces, and (c) in much of the United States). But the blending of individuals from widely disparate research institutes has blended the mix to such an extent that in one large centre I spent a little time at recently there were all three going on around the same table.

It is equally important, if you wish to be trendy and respected, to know whether 'homo' is currently pronounced with the first syllable rhyming with *no* or the French *homme.* You can pronounce the suffice '-ae' as either *eye* or *ee* and the letter -g- can be soft or hard. In a word like 'algae', for instance, you can find:

THE MODULAR PROSE WRITING SYSTEM

MODULE A:

1: it has to be stated that

2: using the arguments elucidated

3: as a resultant implication

4: based on integral subsystem considerations

5: assuming the validity of the given extrapolation

6: with regard to specific goals

7: it may be taken to have been conclusively proved that

8: let us consider as a representative example

9: on the other hand

0: despite appearances to the contrary

MODULE B:

1: the interdependency functional principle

2: access to greater financial and material resources

3: a novel computation involving polyvalent algebra

4: the fully integrated test programme

5: the delineation of specific selection procedures

6: any associated supportive element

7: the primary interrelationship between subsystem technologies

8: the product configuration baseline

9: information presented in future definitive results

0: a large proportion of the interface coordinating procedure

MODULE C:

1: should only serve to add weight to

2: must inevitably take into consideration

3: adds overriding performance constraints to

4: requires detailed systems analysis and trade-off ' studies to arrive at

5: must utilise and be functionally interwoven with

6: refutes the trend of current thinking regarding

7: presents extremely challenges to interesting challenges to

8: recognises the importance of other systems and the necessity for

9: is further compounded when taking into account

0: maximises the probability of program success and minimises the capital expenditure and time required for

MODULE D:

1: divergencies in weak interaction theory.

2: the anticipated fourth-generation equipment.

3: the total system rationale

4: the question of common ality in a conceptual sense.

5: the preliminary qualification limit.

6: any recognisable yet discrete configuration mode.

7: the derivation of specifications over a set time period.

8: the subsystem compatibility testing and quantification routine.

9: all the derived hardware systems.

0: the need to rethink priorities and routine procedures.

Select any random range of four-digit numbers (based on your zip code, the date of your wedding anniversary, or your telephone number, for example). Then pick the numbered phrases from each module. A bstruse yet thoroughly authentic publications can be prepared for speeches, articles or letters of complaint.

al'gî (al guy)
al'jî (al juy)
al*j*e (alje ee)
algi (al gee – hard 'g' as in *goal*)

The significance of trying to sort out your allegiances is not to be underestimated. If you happen to be talking about *homogeneous carcinogenic algae*, and – you never know – they could become a new fashion by a fortnight Thursday, then there are a couple of dozen permutations of pronounciations. The aspiring top-dog knows which one to select every time.

Meanwhile it is worthy of note that in a world where a controversy* exists between tomato and tomato (English and American style), then the utter confusion in specialist terminology is truly amazing.

g) Learn an abstruse style

The expertism that abounds in all nonscientific writing is nothing more than a kind of rhyming slang – it identifies soul-mates, repels all boarders, and confuses anyone with its subtlety of phrase. We have already noted some examples, but for the tyro it is none too easy to see how to get on the bandwagon of obscurity, before it is too late and someone has already worked out what you are *really* saying. If you have now gained experience with the **nomenclature generator** which we saw on page 25, then now is the time to move on to the **modular prose system**, four columns of phrases you should never be without.

One can learn to master the rudiments of a foreign language merely by memorizing lists of set phrases from *My postillion has been struck by lightning* (that old favourite from the Victorian phrase-book) right up to today's *Help, I have been mugged and my credit cards stolen*, so this idea should find a ready market.

If you wish to write to an Expert, or send a wordy communication to one of the organs of Nonscience, even (as an extreme example) to write a meaningless

---

* And how you pronounce *controversy* is one itself.

lengthy letter to the accounts office which keeps sending you an incorrect bill, in the hope it will keep them so preoccupied that they may forget about the matter altogether, then the table is for you.

It may look complex, but in practice it is gratifyingly simple. Suppose you wish, let us say, to query a computerized gas bill and put them off for a few weeks. You carry out the following simple steps:

a) compose a polite introductory sentence;

b) think of a series of four-digit numbers (for example the date of a famous battle, the year of your birth);

c) look them up in the table and write them all out.

This gives you the basis for exactly the kind of letter you need. In the present example we have told our client to set out his telephone dialling code, the year his father was born, and his telephone extension number at work. That gives 0447, the year 1936 and No 5760 which, when read off from the four columns, provide us with our letter:

TO: The Chief Accountant.
Dear Sir,

I wish to query my latest gas bill, which seems to be incorrectly computed due to an elementary programming error in your section.

May I make a few comments which may clear up any confusion? Despite appearances to the contrary, the fully integrated test program requires detailed systems analysis and trade-off studies to arrive at the derivation of specifications over a set time period. It has to be stated that information presented in future definitive results adds overriding performance constraints to any recognisable yet discrete configuration mode. Assuming the validity of the given extrapolation, the primary interrelationship between subsystem technologies is further compounded when taking into account the need to rethink priorities and routing procedures ...

Similarly, if you keep the table by you whilst

154

lecturing you can churn out prose for hours. There is a good three minutes' worth if you merely read the table out as it stands, but with the combination of four lines of ten columns, it would add up to hours!

The principle of using assorted phrases from a random table has an ancient lineage. It was the Franciscan Ramon Llull, who was born in Palma Mallorca in 1234, who first perfected the art that Francis Bacon succinctly described as 'a method of imposture … sprinkling little drops of science about, in such a manner that any pretender may make some show and ostentation of learning'. At one stage he even constructed sets of circular cards which could be rotated to generate random assortments of terms written on each one. He was a far more efficient instructor in the method than I have been in this chapter: all I have included is a sample letter about a gas bill, but Llull once published a selection of 100 church sermons produced by turning his wheels of terms.

Ramon Llull was satirized by Swift in Gulliver's voyage to Laputa. Gulliver narrated a visit to a Professor who had assembled a large framework of wooden cubes on wires. Each wooden block had a word written on it, and by turning a handle endless random combinations could be noted. Any that made sense were carefully copied out and from this works of great importance were compiled. Wrote Swift: 'The most ignorant person at a reasonable charge, and with a little bodily labour, may write books on philosophy, poetry, politics, law, mathematics and theology, without the least assistance from genius.' Little did Swift imagine that by the 1980s this principle would underlie one of the greatest growth industries in the history of the confidence trick.

h) Publish or you're damned
'The only thing that matters is publishing … ' The number of times I have heard that. No-one is interested in *what* you are publishing or even *where* you are publishing it (as long as you do not write for

155

> Rackets on the authorship side of scientific
> publication are normally devices to ensure the
> maximum number of publications, because in
> many environments the number of publica-
> tions is taken as a crude measure of worth.
> Great Men have been known to publish over
> 100 papers per year by affixing their names
> to every paper published by their large teams
> of assistants. Each piece of work gives rise
> to several separate papers, because editors
> prefer short manuscripts.

newspapers or popular magazines: then you are really
in trouble, for that is communicating with the public
and should never, *ever* be contemplated for an instant). Just write out your data with a lot of help from
the modular tables, juggle it around a little from time
to time, and send it round the journals in your
speciality. They will keep publishing it, the libraries
who subscribe will go on paying the inflated prices
they expect to pay for journals, and everyone is most
impressed by it all. Sometimes an Expert copies out a
paper from journals and republishes it in a rival rag
as his own with his name across the top as large as life.
It is even possible to publish exactly the same paper
twice – word for word – with a different title so that
nobody will notice. The record, to my knowldge,
stands at five republished versions, all with totally
different titles.

The golden rule here is to realize why publishers
produce journals. Publishers are tree salesmen. You
can buy growing trees as they stand, in which case
you become a speculator in real estate with a profit
margin of around 20% per year. Next stage is to be a
timber merchant, who cuts the trees and sells them as
wood. The value here is a mark-up of more like 100%.
You can chop up the timber if you wish to become a
wood dealer, in which case the mark-up approaches
the 250% level. If you are sufficiently adventurous you
can produce paper from the chopped-up wood, and
sell that at a value increase of 500%. Some people
print across it and sell it for toilet use (words like
GOVERNMENT PROPERTY are almost a social

156

comment), when the profit over the original goods can eventually total 2000%, weight for weight.

But if you print *lots* of words, especially if they are very *long* words, and bind the pages at the edges, then you have produced a journal. These sell for the most prodigious prices imaginable and the profits reach nearer to 1000% (and sometimes far above that) overall. It is the most cost-effective way to sell trees. Thin authors are there many; but the fat publisher rules the roost and in this form of publishing he is at his very fattest.

The latest trend (now that journals are reproduced by offset litho) is for the publisher to command his authors to provide what they call camera-ready copy. This means that the article is typed (at the author's expense, or at least that of his institute) and the pages are all set out in exactly the final form that will appear on the page. As a result, the publisher has even less to do. Were I to commend an avenue for making a fast buck, then journal publishing would be my choice. That or a gaming saloon filled with one-armed bandits and Pac Man machines.

Here then are some guidelines for aspiring writers of papers. At all stages bear in mind that Nonscientific paper-writing is quite different from authorship. The principle difference is that you must remember not to give anything away.

The title
It is vital that the title gets as much currency as it can. Suppose you were of a mind to investigate the effects of tea on love-making. The boring old scientific way of writing a title for that publication would be something like this:

IMBIBITION OF TEA AND LIBIDO LEVELS IN ADULTS

Now, for an Expert, there are two alternative approaches to the problem. In the one case you are after all the publicity you can gain, so the wording changes to:

157

But it is equally important to know that the paper is going to be mentioned in all the information centres round the world. This includes citation lists, bibliographies, and of course it helps if it appears in as many computer-store data-retrieval systems as possible. These work on a 'key words' basis, where significant words in the title are indexed in a master file. In the first example above, the only words you might expect to see extracted for retrieval would be *tea* and *libido*. Worse still, the second alternative does not add much to this. But look at the title again and put in a good range of key-words, and then you might end up with:

CONSUMPTION OF TERNSTROMIACEAE AND EFFECTS ON ERECTION/EROTIC BEHAVIOUR AND SEXUAL RESPONSE IN ADULT HUMANS RELATED TO AUTO-IMMUNE DISEASE, SOCIAL BEHAVIOUR AND ECONOMIC SUBGROUPS: A FACTOR IN THE PATHOGENESIS OF NEOPLASTIC DISEASE/SCHOLASTIC UNDERATTAINMENT AND MARITAL DISHARMONY?

That would get listed under a mass of headings, including *Ternstroniaceae, erection (penile), eroticism, behaviour, sexuality, responses, disease, autoimmunity, sociology (general), economics (social), pathology, neoplasms = cancer, schools, education, attainment, marriage, marital disharmony, disharmony (marital etc)*. And *consumption,* which should turn up on the tuberculosis files.

This is an enormous improvement of twenty entries over the original two – a simple change of wording puts you in *ten times* the number of retrieval sites. Add in a reference to the nationalities of the people concerned:

ADULT HUMAN BRITISH, AMERICANS FROM THE EAST AND WEST COAST STATES AND WHITE EUROPEANS ON

Then you might find yourself listed geographically under masses of towns, countries, states and counties. This could easily amount to two hundred other entries if the title looked specific enough.

But key words alone do not get the Expert all the media attention he needs before people will accept his ban on tea (or his legalized requirement to drink it by the bucket-full, if that is what he is aiming for). So the aim is then to write a title that does both – it attracts the mass public with its hot-blooded enthusiasm, and gains all the indexing it wants from its broadbased use of key words:

HOW TEA EFFECTS YOUR LOVE LIFE: IMBIBITION LEVELS IN WHITE POPULATIONS OF UNITED STATES, EUROPE AND THE BRITISH ISLES RELATED TO LIBIDO, SCHOLASTIC ATTAINMENT, AND DISEASE LEVELS INCLUDING ARTHRITIS, CANCER, HYPERSEXUALITY, KWOK'S SYNDROME, INCIPIENT HALITOSIS AND INGROWING TOENAILS ...

The possibilities, truly, are many.

In the same way, a title that merely read:

OBSERVATIONS ON CERTAIN AROMATIC AMINE
REACTIONS WITH CAUSTIC SODA

could while away years of contented work, which could all be done by the technician, and would leave you plenty of time for snooker, cards or photography at the lab's expense, would alas be lost in the labyrinthine recesses of the literature. But dressed up it could be a very different matter, heavily cited in every corner of the chemical world:

IONIC MOBILITY AND POTENTIATION IN PHENYL p-TOLUENE-SULPHONATE PRODUCTION FROM p-TOLUENE SULPHONATE/ SCHOTTEN-BAUMANN BENZOYLATION AND SULPHONYLATION BY BENZENESULPHONYL CHLORIDE REPLACEMENT OF BENZOYL CHLORIDE UTILISING LIGHT PETROLEUM

159

That has it all: plenty of key words, neologisms like 'purifactant', and masses of obscure heavy-handedness which will keep people away.

It is always handy to have a publication of this sort hanging on the noticeboard for when the grant committees come around, as it will stop them asking questions. And of course in a case like this the last thing you want is media attention; here the aim is merely for a quiet life and a chance to impress everyone with all those long words.

Introduction

There is absolutely no need here to introduce anything. All you should do is either to repel outsiders (if your desire is to publish something boringly repetitive) or alternatively to make everyone notice your presence. Suppose you have done some work on ultrasound, for example, and you find that it produces changes in cells under experimental circumstances which might affect people in general. Now, there are two possibilities here:

a) you may want to extract as much grant-grabbing publicity from the project, to boost your flagging funds and the status of the laboratory. Or:

b) it may be preferable to prevent people from finding out too much at this stage because it would interfere with the next step of your rather hazardous research project.

In each case you approach the introduction question differently. For (a) you might begin the introduction thus:

Many people believe that ultrasound, commonly used in research investigations, may have long-lasting effects on human cells. Mutations, even cancer, have been mentioned in this respect. We have carried out some important investigations which reveal, for the very first time, the extent of this real risk...

160

It would soon catch the attention of your colleagues and, doubtless, the media too. Here now is the same idea exactly but couched in the more circumspect terms you might adopt if you had a dangerous project under way and did not want people to stop you going too far:

Intracellular disruptive microcavitation phenomena associated with the diagnostic and experimental applications of ultrasonic irradiation in the paramedical and biologic disciplines with its correlated proclivity for heredito-genetic sequelae and intranuclear degenerative manifestations in relationship to transformed cell replication and mutative disorders ...

That takes us only about one-third of a sentence into the earlier version and, though it gives all the leads your fellows will want to have to follow up your dastardly plans, no-one else will ever latch on. Your success depends not on the work you have been doing, but on how it gets to look from the outsider's point of view.

Methods
If you have a new method, for heaven's sake do not let the fact emerge here. Keep it to yourself, as you will need it for the next couple of decades whilst the grants build up waiting for your ultimate breakthrough. So skate around anything that might be of value to an enemy.

Instead, bolster up the methods with lots of impressive makers' names, temperatures quoted to small fractions of a degree (expressed as decimals, of course) and tedious repetition of the minutiae of experimental detail that other people will already know about. I know of individuals who list the kinds of filter-paper they use, give analyses of glassware, weights of superfluous components, dates, times ... you name it, they print it.

They end up with a long and detailed paper which gives nothing, at all, away.

Observations
Take care to reveal nothing here.

161

Results

In this all-important section you give the impression of reaching your grand conclusion. In fact this is the part you should write first. The best approach to this section is to carry out all the calculations you require to give you the ideal results. It is these you publish. Then you work backwards until you come to the question. Do not try to do it the other way round as it rarely works and it is accurate results that people need in this world. The classical example here was Gregor Mendel, the famed monk of Brno, who in the latter part of the nineteenth century set down the basis of today's genetics. He did it all by fiddling the results he expected to get with his sweet peas, and then writing the experiments later. It was not until the time of the Second World War that people began to repeat his work, and found that the results could not possibly come out as well and accurately as Mendel insisted they did. It then turned out that this was because he had made them all up as he went along.

Discussion

Experts, having horizons that are limited to the breadth of their minds, cannot write discussions. Some merely leave out this section. Others copy a piece of prose from some earlier publication and change around the words a bit so that people will not notice. Another possibility is to write out alternate sentences from the *Introduction* and the *Results*, with occasional phrases to alter the flow from the original.

However the *Discussion* is where you should make all your excuses. If you feel that the *Results* are wrong, then mention that they are reckoned to be within 'normal experimental error' or perhaps that they are correct 'with one order of magnitude'. If you find you have missed something out, then say it was disregarded because of its elementary nature. You may eliminate experiments that did not fit your own prognostications, for instance, by saying that 'several sample runs were subjected to further analysis', or you can cover up a mass of uncertainty by the liberal use of

162

disclaimers including 'it is widely accepted that ... '
or 'it is generally conceded ... ' even if it isn't and you
merely wish it was.

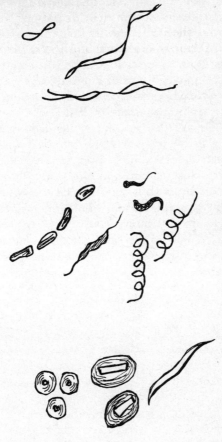

Three examples of diagrams of bacteria culled from standard
reference works. These are organisms invented specially for
textbooks: nothing like them exists in the real world.

Illustrations
Here you find ample room for a) confusion, or b)
impressing people. Do not try to illuminate matters
through illustrations. For example, from time to time
you come across diagrams of microorganisms, say,
which give a reasonably good idea of what they look
like. Now that is no use to anyone. What you want

instead is a distorted and vague appearance that looks like nothing on earth. This keeps the curious at bay and guards your own secrets with a mask of mystery. It is important to study the microorganisms that have been invented to illustrate books. Almost any reference work will show them in their prime. They are quite unlike any organisms that actually exist.

Diagrams and graphs should be cluttered with facts and figures, or fictions and figures if necessary. All possible avenues of clarity should be eliminated. If you are trying to plot results on a graph it is quite in order to obtain special graph paper with exaggerated axes on a geometric basis which may be logarithmic or based on squares and cubes, or indeed you could work out some strange combination for yourself. In this fashion almost any data can be coaxed into whatever shape of graph you are aiming for, whether it is a neat gaussian distribution, an asymptotic curve, or a straight line. Histograms are fun too, so long as they are made to look adequately confusing. And this brings us to the correct use of mathematics in a publication.

The acceptability of research
It has been represented in the foregoing paragraphs that for the results of research to be incorporated in criminal policy they have to be both conceptually acceptable and to some extent consonant with information received from other sources. This process[7] may be illustrated diagrammatically:

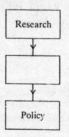

Research and Criminal Policy by John Croft (Home Office Research Study)

Maths, Data, Figures
Graphs are all very well, but there comes a time when you have to quote some genuine figures (by which I mean that they really are figures, that is all, and not

that they mean anything genuine themselves, which they don't). The first art to master is the means available for making figures that are not precise look as though they are.

We are helped here by the notion of *degrees of accuracy*. In the old days you could use traditional measurements to give a fair idea how accurately you were measuing any given quantity. 'A bucketful', for example, was a clear enough amount and it was fairly obvious you were not measuring that to the nearest milligram. But if the bucket holds two litres then you say it contains '2000 ml' and that does not mean 'a bucketful' at all. That means two thousand millilitres, no more, no less, which at once implies that it is to the nearest millilitre that you have been measuring.

Again, suppose you see a man coming into your section and you measure him. He is a six footer. You would in the old days write that up as 'a six-foot man' which, if he was not some freak over-endowed in the appendage sector, gave an idea of his size and also of the degree of accuracy with which you had assessed it.

But now let us bring that up to date. We now measure metrically, and to the millimetre. So the six feet becomes exactly 1,828.8mm (or 1 828,8mm or even 1828.8mm depending on how you view these things), which is the exact equivalent of six feet in metric units and has to be quoted in full. The datum presented in this new format implies a degree of accuracy down to the nearest 100μ (or 100 μm as that would now be, of course) which is patently absurd, but absolutely necessary.

In just this way you can provide the most empty-headed reasoning with a veneer of authority. You can express guessed speeds to the nearest centimetre per second, estimated wage bills with the nearest portion of a cent, and biological data down to the minutest fraction of a gram, a millimetre or an Angstrom unit instead.*

* My apologies, the Angstrom unit has since been banned, but the principle is the same.

Whereas you might once have said 'three dozen', meaning a number somewhere between 30 and 40, it is these days necessary to go for definitive accuracy (even if it is practically meaningless to do so). And suppose you really do want to express the same kind of approximation? Then the procedure is as follows:

a) select the figure to which you wish to allude, say 1000.

b) quote it as something extremely accurate which is around the 1000 mark, e.g. 987.56.

c) then make it look approximate once again EITHER by quoting the number to the nearest ten, OR by suggesting it could easily be ten times bigger or ten times smaller. The alternatives are as follows:

> i: 987.56 ± 19.5
> ii: 987.56 correct to one order of magnitude

The first of those could apply to any figure between 986.06 and 1007.06, in other words it means 'a thousand, approximately'.

The second alternative covers the range 98.756 right up to 9875.60 and so it means 'a thousand, *very* approximately'.

Finally, the technique of working out numbers as an average. If you were to claim in some public-house argument that most people make love twice a week then (even if you won the argument) 'twice' is a pretty ordinary-looking amount to be arguing over. What you should do is take a selection of claims and then strike an average (which only non-Experts refer to as an average; in the trade they are always a *mean*). This gives you something with decimals to play with, thus: '*The average couple make love 2.64283 times per week.*' Few will even begin to argue with that. It sounds accurate, looks impressive, and has all those authoritative figures to prove the point. The layman is mesmerized by data.

There are several averages to choose from, among them the arithmetic mean, the weighted arithmetic, the geometric and the harmonic. All look impressive, and mean little to anyone not well versed in the artificialities of mathematics.

That concept – artificiality – should be more widely understood. We are raised on a diet of one plus one equals two, and soon find ourselves thinking that mathematics has a self-evident wholeness that is somehow superior to nature itself. However, that is only because mathematicians restrict themselves to sets of circumstances where mathematics does work. If you wish to experience something of the arbitrariness of mathematics, then here are some interesting problems.

1) If it takes one man one day to decorate a bedroom and another man two days, how long do they take to decorate the room together?

Obviously, you would think, the first man only has to do half the work, so that means that his labours take half a day. The other worker has half to do too, so his time is halved too. Thus, it takes a total of $1\frac{1}{2}$ days to complete. But that is longer than the first man would take on his own ... think about it.

2) One cup of popcorn plus one cup of milk makes, what, *two* cups?

No, one.

3) If one Mona Lisa costs $5m, how much do three Mona Lisas cost?

Acknowledgements
Do not get carried away on a wave of untypical honesty. One example from the *Journal of Biological Chemistry* read:

I wish to thank Dr Lester who not only suggested most of the experiments herein but greatly helped in their interpretation, and A. S. Bottorf and B. Fravel for their excellent assistance in performing these experiments.

That almost suggests to the cynical reader that the experimenter did next to nothing himself. It would have been better by far to go for something less admirably self-effacing, like:

I am grateful to Dr Lester for suggestions and to Messrs Bottorf and Fravel for technical help.

Do not put too much responsibility onto others in case they object. But on the other hand it is perfectly normal to acknowledge help from eminent names, even if you only said 'Hi!' to them at a conference, and all they said back was, 'Yes, it is, isn't it?' as they stepped from the elevator on the 17th floor. They will not remember having never met you, really, and will be flattered at finding another mention to go into their own bibliographical compilations. So throw away a little line like:

The author is grateful for discussions of certain aspects of this work with Lord Colloid and Professor E Coli of the Institute.

No-one will ever know the truth.

References
The table of references at the end of a paper is best compiled in the now time-honoured fashion of copying out the longest bibliography you can find and adding a few items of your own to make it that much longer. It is important to work out who is going to see the thing. If it is going to be read by Lord Colloid, and he hates Dr Coli with a rare degree of feudal bitterness, then it is no good referring to several of Ed Coli's papers in your bibliography and leaving out all mention of Arthur Colloid. The reverse would apply, of course, if Ed was going to review your paper instead of Arthur.

I recall a paper in which reference was missed out to work done by a close friend of the referee. Back it came with a warm little note adding: 'You really ought to mention this seminal work.' The change was made, but in the pressure of work the paper was a month or more in reprocessing. By then the referees had changed and the new lot couldn't stand the sight of this peerless individual. The note was sent out as before, this time recommending the same section be

168

taken out: 'There is really little point in referring to this superfluous item,' it said.

And a British microbiologist not long ago wryly produced two rejection slips from rival journals, to each of which a paper had been submitted (it was the same paper in each case, which, though strictly against the rules, is a perfectly normal practice). One said that the paper was 'rather superficial and a good deal too preliminary in such a new field,' whilst the other said it was 'unoriginal and only describes work already familiar'.

The mainstay of a successful publication is a lengthy bibliography. A good paper with few references stands no chance at all, but a poor paper with a list of references as long as your arm (or its metric equivalent) will have you in there, publishing faster than any rival.

A Precaution

There is one final possibility that should be mentioned. It has all the Nonscientific bravado of the cult of the Expert, but is ordinarily too outrageous to be worth risking. In spite of this, it has been tried in the past and (although it could never be recommended) an Expert fearless of all consequences just might be tempted to try it again.

What you do, in a gesture of breathtaking simplicity, is to play two cards at once: *you publish two papers in separate journals each of which argues that the other is false.*

K.O. Kutschke, of the Canadian National Research Council, who is Editor of the *Canadian Journal of Chemistry*, says that in February 1978 his journal received for publication a manuscript called "Influence of electric charges on the corrosion of metals" a paper previously published by a group of Swedish authors — J.A. Hedvall, Nils-Gosta Vannerberg and P.O. Blomqvist

The Editorial Board of the *Japanese Journal of Medical Sciences and Biology* here announces retraction of the following paper from the Journal:

"Effect of Platinum Compounds on Murine Lymphocyte Mitogenesis" by E.A.K. Alsabti, O.N. Ghalib, and M.H. Salem. *Japan. J. med. Sci. Biol.* 32(2), 53-65, 1979.

Judging from several lines of evidence (1–5), this work must have been authentically carried out by Drs D. Wierda, and T.L. Pazdernik, whose paper was almost simultaneously published in the *Eur. J. Cancer* (15 (8), 1013-1023, 1979) under the title of "Suppression of Spleen Lymphocyte Mitogenesis in Mice Injected with Platinum Compounds".

In that form it seems too obvious, too simple to work. The advantages are clear enough, though: whichever point of view eventually becomes fashionable, you could always point to the paper which supported that thesis and confidently proclaim yourself to have thought of it first all along.

The disadvantage is being caught out. Even so, you can find precedents in the literature. Serious students may like to scan the pages of the prestigious *New England Journal of Medicine* for 18 May 1978,* where the editorial and the letters raged about two papers which had been published by the same people in January 1978, each of which argued that the findings in the other were wrong!

The two papers were on mono-amine oxidase (MAO), an enzyme, found in the blood platelets, and its relationship to schizophrenia. One paper was entitled *Platelet monoamine oxidase in chronic schizophrenic patients* and it concluded that a comparison between the platelets of paranoid schizophrenic patients and non-paranoid schizophrenics showed there were 'no significant differences' between the two groups' MAO levels. The other publication, entitled *Are paranoid schizophrenics biologically different from other schizophrenics?* concluded exactly the converse, namely that there *were* significant differences between paranoid and non-paranoid schizophrenics' platelet MAO levels.

In each case, the papers shared two authors, Dennis L. Murphy and Richard J. Wyatt. The *New England Journal of Medicine* fumed: 'We are puzzled ... by the virtually simultaneous publication of two apparently contradictory papers, one in the *Journal* and the other in the *American Journal of Psychiatry*. Despite the fact that these papers share two co-authors in common, neither manuscript, as submitted, referred to the existence of the other.'

When the issue was first raised, the authors were

* Psychiatric papers appeared in:
New England Journal of Medicine, 298: 61-66, 1978
——————————— , 298: 1150 *et passim*, 1978
American Journal of Psychiatry, 135: 95-99, 1978

quick to retort: 'It has been our policy not to refer to unpublished data in a published paper,' which is an ideal excuse well worth remembering. There is nothing to stop anyone from adopting as their policy the refusal to countenance any unpublished work. And if you are publishing two papers it is obviously possible to decline to mention each in the other because they are both unpublished at the time.

The critics wrote: 'To dismiss one's own discrepant results as being "unpublished data" and therefore not open to comment defies common sense and is, to say the least, disingenuous.'

All I can say is that the editors of the *New England Journal of Medicine* ought to be more circumspect in their criticisms. In this era of the cult of the Expert it is beginning to look as though what that journal dubs as 'disingenuous' is actually a foremost principle of behaviour. In the contemporary interpretation, people ought to get honorary professorships, civic awards or a major prize for that kind of stunt.

Experts of Imagination

It will by now come as no surprise at all to learn that the literature has thrown up several examples of Experts who had never existed.

The two sociological assistants referred to by Burt, Miss Howard and Miss Conway, seem to have been figments of his fertile brain. But at least 'they' were only helpers. Far more eminent was the legendary Dr O. Uplavici (1887-1938) who played a leading role in the development of parasitology around the time of the First World War. He was cited in a host of bibliographies. Research workers heard of correspondence with the great man, and of the discussions in which he had engaged. For years he figured as an influential, behind-the-scenes figure in the medical world.

It was not until 1938 that it was suddenly revealed that Dr. Uplavici had never existed. What happened was that a prominent Czech physician at the University of Prague, Dr. Jaroslav Hlava, had carried out some experimental work on dysentery. He was in a

171

hurry to publish, and so decided on a preliminary account which duly appeared in the *Journal of Czech Physicians* for 1887. The title, in English, was **On Dysentery. Preliminary communication.** In Czech, this was **O úplavici. Predbezne sdeleni.**

The topic – O uplavici – became listed instead as the author's name – Uplavici, O. Within a few years, *both* Uplavici and the original author Hlava were being listed as research workers in the dysentery field. Later still the name was indexed in the *Index-catalog of Medical and Veterinary Zoology,* included in Part 10 for 1905. The accolade came later: in Part 31 (1910) he has been given a doctorate, for his name then appeared as 'O. Uplavici [Dr] '. It was a fitting climax to a career that was distinguished by never having taken place.

More recently, in 1972, a mythical psychiatrist named Myron L. Fox was created by an actor who went to a meeting and read a paper about **Mathematical Game Theory as Applied to Physical Education.** After the paper he was questioned by the audience, in the normal way. The man knew nothing in the world about psychiatry, nor about game theory, least of all about physicians or how they might be educated, but bluffed in meaningless prose (of the kind you saw on p 152) and answered using this.

In a questionnaire, all the specialists in the audience loved him, not a single person gave more adverse reaction than positive support in the analysis. The questions showed him, not to be a fraud and a charlatan, but to be 'knowledgeable', and his answers were a 'good analysis of the subject'.

CHAPTER TEN

Experts and the Public

The spreading acne of Nonscience has taken the new and wild-eyed race of Experts into every walk of life, driving the bewildered victims like withered leaves before a plastic brush in Autumn. It is suprising how seriously they are taken; but there is a self-centred headiness, almost an ebullient mania of oppressive righteousness, about the Expert of today. He may have done nothing more than classify a group of wayward plankton of no significance to anyone but their mates, but he intimidates all outsiders with the piercing eye, the jutting chin, the uncompromising manner which demolishes all opposition.

Experts always tell you things, they never ask. Every Expert pushes aside criticism or probing, he merely asserts. Experts fly from ideas like midges from fire-smoke; they congregate around piles of data instead. Above all, they adore their power to rise above everyone else in the scramble for prestige, without the slightest wisdom, or worldliness, or even common-sense, behind them.

Experts pontificate about things like the weather (about which nobody knows much) or earthquake prediction (about which nobody knows anything – and it would make little practical difference even if they did). After all, just to take this latter example, we know a lot about where earthquakes occur and how low buildings can minimize damage, but that does not interfere with idiots building the largest buildings across the most dangerous fault of them all and living there contentedly watching the old seismometer swing.

People know how dangerous cigarettes are, but that fact does little to stop them smoking. They know how violently destructive TV is, but it never alters their obsessive addiction to staring at it; they know how large cars consume the earth's resources, pollute the air and damage their health, but still brandish them like tokens of materialism.

Yet Experts have done much to keep us from truths that might worry us. We have been taught to look back in horror to the bad old days when people used to drop like flies from diseases such as diphtheria and the plague. Yet we have our fellows dropping from coronaries, strokes, obesity and the rest (which will shock our descendants just as much). We know about those miraculous antibiotics and thank our lucky stars that they are around for us, even if they were unavailable to our grandparents who used to get so many terrible diseases. Yet this is muddle-headed, too. The great killer diseases caused by infectious bacteria were largely banished long before the antibiotic era by improved standards of sewerage and better hygiene. In any event, we have no antibiotic that works against viruses and how few Experts mention that.

Perhaps the cult of the Expert is a movement designed to occupy our waking hours and prevent us from becoming worried by too many major decisions. How else could you explain the undemocratic aloofness of the planners who wildly (and expensively) tear down well-built old centres and erect sagging new ones in which everything goes wrong? Or the mystery of the Experts who alter around the values people feel secure with? Is this why Cape Canaveral became Cape Kennedy just long enough to get the world used to the idea (with changed maps and all) before being switched dramatically back to Cape Canaveral again?

Was it not an example of Nonscientific thinking that gave the world monetarism, in which everything becomes transferred to monetary value? This has been the greatest single cause of moral decline and civic discontent since the industrial revolution. Today you do not perform tasks because of their merit, but because of their pecuniary value; you judge people not by their worth as individuals, but by their hold on

174

the capital market. Once financial value comes in, the old considerations of compassion, kindness and altruism are instantly out-moded. So perhaps this is the answer: some Expert has provided monetarism as a final nail in the coffin of morality. If so, fear not. Like other fashions it cannot last, and within a few years it may be the bubble-gum economy or bodily-friction-as-an-energy-source which preoccupies us instead.

It is not only in the noisy corridors of legislative power that there lurks the Expert mentality. In the subtler and more secret recesses of the social service industry there are plenty of these manic, mindless dictators busily planning tomorrows for their clients. Their power is incalculable. Few people realize that social services officials have the power to confine people against their will. Young people can be detained in custody. And if an ill-advised parent allows a child to enter care temporarily, then it is never so easy to get the child back again. A 1981 report from Sweden gave some astonishingly vivid examples, which can be matched with similar case-histories from elsewhere around the world. Swedish law allows any child over ten months to go into a day-care centre so that the mother can resume her work. One mother declined to do this, and her child was taken into care because, it was said, the mother might become over-protective. Another mother refused to let her children wear jeans to school. The children, both girls, were teased about it, and, when their mother encouraged one of them to study the violin and the other ballet, the girls were taken into care by a posse of officials and police who read out the legal documents committing the girls in front of the entire class, before dragging the girls away.

In another case history, a child was taken into care because of an anonymous report from a malicious neighbour that the mother was a secret drinker. After investigation, it transpired she was a devout tee-totaller, and loved her children very much. This had a disastrous effect. It is not in the nature of any Expert to admit he may have been wrong, and so the

175

grounds for detaining the child were changed. The reason now became 'because the mother and child are too close for comfort'. Meanwhile another girl of 15, who had her ears pierced in spite of her parents' wishes, was seen crying about it in school. 'My parents say I must stay in after homework,' she said. By nightfall she was securely in State care. The parents were not even consulted.

Perhaps the worst case is that of Solbritt Lakkampal, herself a day-care school supervisor. She once asked social workers for advice because of family rows, and when she found the advice was useless she told them so. They called back to see her some time later, and she sent them packing for the second time. When they returned, she was on the point of leaving the house with her new-born baby. They locked her in a room, snatched the child and for the following two years she has been trying to get the child back. Lennart Hane, a Stockholm lawyer, claims there are 23,000 children in care in Sweden. That is seven times as many as in Britain, for example. Of that total it is claimed that only about 500 need to be in care at all. But it all provides security of tenure for the Experts, if nothing else.

And this is by no means the worst kind of activity that you find in the social sciences. As Margaret Mead once said, in a little footnote, though it would have done well in capitals, you can even have the effect 'in which a medical school or a discipline comes to be regarded as a place where animals are mercilessly tortured, and human subjects are manipulated and brain-washed'. That is close to the situation we are now fast approaching.

There have been many examples of experiments involving humans, of which perhaps the nastiest examples concern the use of surgical mutilation of the brain. It does not appear necessary to conclude that this is the limit of what man can do to man in the name of research. The interesting experiments of Professor Stanley Milgram showed how far people in this era of the cult of the Expert can be induced to go.

In the early 1960s, Milgram decided to investigate the extent to which people would obey an authoritative figure with an unquestioning manner and a white coat (whom we recognize as an Expert), even if the instructions involved cruelty to another human being. An advertisement was placed in the local press, proclaiming:

PUBLIC ANNOUNCEMENT: WE WILL PAY YOU $4.00 FOR ONE HOUR OF YOUR TIME – PERSONS NEEDED FOR A STUDY OF MEMORY.

Age limits were 20 to 50, and no students from high schools or colleges were accepted.

The volunteers found that they were scheduled for joint appointments, two at a time. They were given a brief run-down on the experiments. They were told that little was known about learning (this is true now as it was then, of course; we know absolutely nothing at all about how the human mind functions, in spite of what it is popular to claim).

The two people were then asked if either had a preference to act as **teacher** or **learner** in the trials: the decision was often taken by drawing lots. Then the **teacher** would be seated at a control console, whilst the **learner** would be led into an adjoining studio and sat in a kind of electric chair through which small electrical shocks could be administered. In most cases the volunteer playing the part of **teacher** asked in advance if the shocks were painful. The answer was: 'Although they can be extremely painful, they cause no permanent tissue damage.' With that, the tests began.

Each **learner** was asked to pair words, an adjective and a noun, such as *blue box, wild duck, nice day* etc. He was shown an adjective with a choice of nouns, in this format:

Blue: sky ink box lamp

The **learner** then communicated his choice by pressing one of four buttons in front of where he was strapped into the electric chair. This lit up a light in a panel in front of the **teacher** volunteer, who was also

177

able to listen to the subject on a closed-circuit loud-speaker system.

In front of the panel was a set of 30 switches labelled with voltages from 15 to 450 volts. There were also a set of labels ranging from SLIGHT SHOCK to EXTREME INTENSITY SHOCK: DANGER with two switches after the highest verbal warning labelled baldly XXX. The **teacher** volunteer in control was told to move to one level higher each time the subject did not give the approved pairing or made a mistake. The supervisor told the volunteer, if he expressed a dislike of the experiment and wanted to let the **learner** out of the chair, 'You must go on until he has learned the words correctly, so please continue.'

When the results were published in 1974 Milgram admits he was astonished at the number of people who were willing to administer even very high voltage shocks which could clearly have come close to killing their partner.

What the volunteers did not know, however, was that the **learner** was always the same – an actor named James McDonough, a chubby and cheerful man who played his part with complete conviction. When the interviews were held, the genuine volunteer was given the chance to opt to be the **teacher**. If he did, he was allowed to proceed; whereas if he didn't the interviewer produced cards and told the two to draw lots for the task. Actually, *both* cards bore the words **teacher** so the volunteer was bound to end up choosing that rôle. Each time that happened, McDonough would look at his own card, say: 'Uhu, so I'm to be the learner' or something of the sort, and would march cheerfully off towards the experimental room.

His cries and screams of agony were extremely convincing, says the report of the experiment. Many of the volunteers asked if they could stop now, but a very high number went on when told to do so. They all knew – or *thought* they knew – that the **learner** was genuinely getting the shocks because at the beginning

178

of the test they were given a 45 volt shock themselves to find out what it was like. The conclusion was clear: that ordinary men and women will risk the lives of each other if told to do so by an Expert in his white coat.

The garb matters, for it became clear that when a person (described as an ordinary man in the text) dressed in casual clothes gave the orders, there was less conformity from the volunteer than if the supervisor had on a white coat. In many ways this is the ultimate extension of the kind of regimentation we have seen in this book. People learn at school that obedience is the most acceptable societal response to authorities.

At the end of it all people become so conditioned that they will even threaten each others' lives – if told to by that figure from the cult of the Expert.

Electronic Brains versus Human Bodies

I do not doubt that one of the secrets of success behind Professor Milgram's hearty experiments was the computer-like array of lights, buttons and switches with which the volunteers were confronted. Ordinary people have an ability to believe anything of electronics. At the present time anything with a micro-chip is bound to be miraculous, just as any figure that has been (and this phrase, to be authentic, should really be spoken in a Southern Californian accent, breathless from tennis, jogging or juggling) *processed through a computer facility.*

I remember as a child of eight seeing a TV programme about a device controlled by electronics which looked like an aluminium tortoise. It trundled across the floor, sensing objects in its path and turning away, and was said by its developer to be at the same level of intelligence as a two-celled organism. At the time it puzzled me. Little organisms, two cells or no, could surely do more than trundle.

Today the widespread talk is of super-intelligent machines that can challenge human intelligence. Indeed if the headlines were to scream that a new

179

computer has been made that is ten times more intelligent than Einstein (or perhaps even a zillion times) it is likely that most people would believe it. Facts are different, however. No-one has the least understanding of the workings of the human brain, and for all the ability of electronic machines to store and process scraps of data very quickly indeed and on a scale that would put most filing-cabinets to shame, there is no computer which shows any degree of intelligence at all, and certainly nothing that matches the mind of a microbe, let alone a man.

That is not the over-statement it may appear. If you look at a single celled organism like a protozoan, for instance, you can see it search for food, reproduce, avoid unpleasant stimuli and change its behaviour in a manner that is astonishingly complicated. It has long delighted me to point out that, since brains are composed of cells, and this organism is only one cell itself, it cannot have a brain; yet it succeeds in doing, in its own terms, most of the things that you or I can do. There is no hope of producing any computer system which can simulate those actions, let alone make something which could challenge them.

One way in which the computer has been publicized as a threat to humans is by claiming that automated devices are taking jobs from the masses, which in turn limits employment prospects. That is nonsense. If it were so, then the unemployed would be the clerical and professional people whose methods were being superseded by the machines. In fact, two-thirds of the unemployed are in the unskilled classes, and that now includes a high percentage of intelligent but uneducated teenage school-leavers.

Also there were plenty of people out of work from the professionally qualified classes in the 1960s, when few if any of their jobs might have been 'threatened' by computers built with micro-chips because there were no micro-chips available to build them with. I tossed a little rock into the serenity of the 'white-hot technology' era of the 1960s with a survey of such people, and came across people with excellent academic qualifications who were working as carpet fitters, a car-washer, barman, bread van driver, ward orderly and petrol pump attendant; three labourers (one each digging trenches, working for the railway and on a farm respectively), and a few stranger ones which included an assistant rocking-horse maker and a stamp dealer's mate.

This was a surprising finding at that time, and I have suggested that we should try to analyse the qualification gap that the situation reveals. Not that we ought to educate for a specific career, far from it; the 'relevance in education' debate has often become trapped in that cul-de-sac in the 13 years since my first qualification gap article. What we ought to be doing is giving full expression to the independence and vitality of people as individuals, and not indoctrinating them with the received values of an arbitrarily materialistic society. In some ways there ought to be some kind of movement or a campaign to close that gap between the brain-dulling excesses of modern training schemes and the need for people to fulfil themselves.

However, the cult of the Expert can help us put that into an entirely different light. The obedience training can be seen in this way as a part of obtaining Experts for tomorrow, rather than as the blunting of the mind. Those long queues of misfits turn out to be the raw material for the Expert-dominated society into which we are sinking, and every further twist of the tale, every change of standard, each ban, regulation or confusing new rule, becomes a harmless way to pass the time.

In the bureaucratic societies official documents are always handled by invisible people. (A passport goes through a slot in a booth, and comes out again some time later; the fact that you cannot see the individual is a deliberate policy decision intended to make the visitor feel subjugated.) In our Western bureaucracies, unseen Experts make life-and-death decisions, based on flimsy evidence, which the public accept.

If a figure is plainly erroneous, or a decision

181

self-evidently silly, it is much less likely to be challenged if it has gone through its own 'slot in the wall' – and has been computer processed. The old-fashioned term for a computer was *electronic brain*, though 'electronic moron' quickly arose as a more truthful alternative. Computers are exceedingly unintelligent, indeed I have yet to meet the *Amoeba* who would consider a computer any kind of equal.

Here is an example. Suppose you set a problem for a child in the form: What is sixteen times seven? There are three things the child might do:

a) he could tap out the results on a calculator, which is a neat way of never having to do the sum at all;

b) he could multiply the total in seconds by using multiplication like this:

16
7 x

112

c) he could out down the sum in this form:

7
7 +

14

and repeat that adding 7 on to the total 16 times.

Answer (a) is no answer at all, for the child is merely telling you what figures appear on a dial. That is the answer which the Expert would like to see, of course, for it shows reliance on the appurtenances of technology rather than the minds of people.

The alternative (b) is the quickest method, for it gives an indication that the child (along with learning to walk, to eat, to dress, to speak) has also learned to multiply and is prepared for the innumerable occasions when that helpful ability is necessary to solve a problem.

The third (c) is the plodding and long-winded process of simple addition which you would need if all else failed.

Of the two last alternatives, it might be fair to say that the multiplication-sum approach was the intelligent one, whilst the tedious process of repeated addition was laborious and wasteful of time which could be spent in more enjoyable pursuits.

But that is how computers work out their problems. It is the fact that their circuitry enables the accumulated calculation to be completed in the flick of an eye and the flash of an electron that gives them the basic attribute of rapidity. But rapid is all they are; not intelligent. In other respects we are not so easily overwhelmed by mere speed (a charge of explosive could demolish a building quicker than a team of workers, for example, without anyone imagining that speed means nitroglycerine is hyperintelligent).

And that is how it is that the humming circuits, warbling loudspeakers and whirring tape-readers of a babbling computer have come to impress us. They stand in locked rooms scented with scorched dust and ozone, impressing everyone with attributes they do not possess. The mere fact that (to use the phrase so often quoted by Experts in the field) 'computers can do things no human brain can do' is nothing to get excited about. So can a butterfly or a worm. So can a pin, a matchstick or a paperclip, and there is nothing unfathomably remarkable about that.

If a bank teller fails to provide you with money you legitimately require, you can argue with him or write to the manager. But what happens when a computer-operated cash dispenser instructs you to insert your Personal Identification Number, and then, after a few seconds of muted ticking, flashes out an illuminated message reading CARD RETAINED? Why, you meekly walk away and wait for the card to home in on you through the post in its own good time.

There are many people who have reacted against the computer myth by exploiting its characterless bludgeoning for their own purposes. In many countries magnetic codes are used on bank paying-in slips. One ruse that has been resorted to in the United States was for hard-up customers to go into a local bank and leave a few of their own coded paying-in slips on the document shelf with the bank's own supply for irregular customers. People who had money to pay in, and who did not have their own book with them, would pick up a slip from the shelf, complete it with their own details, and hand it over. The computer would sort according to the invisible magnetic code, and random sums of money were in this way diverted into the desired account.

It takes time for sums to be debited to the master record in the bank headquarters, and there are many enterprising youngsters who have gone into a bank to open an account with, say £100, and have then called back around lunchtime and drawn out £90 of it. When the staff have changed round, the customer goes in and draws out another £90. Because of the delay in processing the debit, the clerk's enquiry of the account file will display the full £100 still intact. With care, people claim to have got away with that time after time, indeed one Iranian student left for home with his suitcase filled with money which he said he had obtained in exactly this way.

In Germany computer programmers experimented by putting a 'repeat' function into a routine, which issued them with hundreds of identical pay cheques. That scheme was not uncovered until one of the conspirators tried to cash a dozen of them at the same time. And nobody knows how much money is filched from accounts by the use of rounding-down accumulations. This procedure is involved in accountancy when there are large numbers of tiny sums of money. Many of the programmes disregard the pennies and round sums down to make the calculations easier. The small sums can be collected in a subroutine devised by the operator for his own benefit. It is said that there are many people doing this at the moment.

The only example I know which came to light was when a public relations company decided to take the first and last name on an employee list as sample material for a survey. It turned out that the last man on the list was a Mr Zwana, whom no-one could trace. It seemed he had a large sum in his account, all collected from occasional pennies ... The name had been inserted by a computer programmer who devised it to be last on the alphabetical list, and then simply commanded the computer to accumulate all the extra sums in this last account of all.

One pastime enjoyed a considerable currency under the name of 'computer roulette'. The idea was to cut, with a razor blade, several extra punched rectangles on an accounts card received in the post. There were many claims made for the results: one was that an expensive magazine subscription was indefinitely extended; another was that a series of monthly demands was automatically terminated.

The results are easier to predict if you understand the code used in the punched card system, and the meaning of those anonymous little holes is familiar to you. Using this knowledge one defecting Expert filched £50,000 out of one of the largest catering firms in Britain, and an American speculator was able to change his bank overdraft limit from $200 to $200,000 with the aid of a cutting knife and a steady hand.

People do this partly out of greed, no doubt, but also because of a sense of grievance against computers. But this is not the right target – computers are only processing machines, and it is the people who run them who make the mistakes, and the vogue for believing in the whole system which compounds them. In that sense, a profound revulsion for the cult is entirely understandable.

Some of the problems are irritants, caused by thoughtless design of a computer programme. A plan for the City of Worcester designed by a complex computer system positioned a new trunk road on the basis of detailed maps and traffic projections, only to site the new throughway right down the nave of the cathedral. In London an excessive electricity bill was sent to an old lady who killed herself (the sum due had been exaggerated in order to force her to arrange a proper meter reading, as she had been sent estimated bills for several quarters). When a fault caused another high bill to be sent out to a customer, he was told he should pay the sum due and then wait until he next owed them something, as there was no routine available for making refunds out of the central accounting system.

The development of a device which could monitor the levels of hormones in the bloodstream of mothers-to-be at term gave rise to a huge explosion in induced births during the late 1970s. The idea was publicized as being the procurement of reliable and safe parturition, but this was patently untrue. Induced labour was known to be more difficult and more demanding for the mother, and it caused a raised incidence of side-effects (including damage to the uterine wall). But it did confer several distinct advantages on the hospitals where the system was used. First of these was the fact that the births could be arranged on a nine-to-five basis, which made the working day more routine for the medical staff and prevented them having their off-duty time interfered with, and secondly it founded yet another branch of Nonscientific high technology, with all the financial benefits and status that this brings with it.

The mothers suffered in consequence, but little attention was ever paid to that.

There is a fundamental idea that runs through the course of the cult of the Expert. It is that if something *can* be done, it automatically *should* be done. In this example, the existence of birth-induction machinery was taken to mean that from now on all births which might be induced, must be induced. In my view this is as erroneous as it is widespread. Mankind has evolved with certain strictures and with a complex range of inbuilt responses, and the aim of our modern life ought to be to act in sympathy with them wherever possible. I believe it is not right to interfere with any naturally effective system, unless you *have* to.

Viewed in this light, birth induction should only be used when the baby or the mother would suffer without it. In every case where it is not absolutely

186

necessary, it should not be involved.

The difference this principle of 'necessity' makes in practice is considerable. Let us suppose you have a unit where there are 350 births per month, 5% of them needing induction for medical reasons and 10% of them being totally unsuited for induction. Using the received values of the cult of the Expert, you would calculate that only the 'non-inducible' category would be excluded from the programme of induced births. That means 90% of the births are induced artificially, which on the figures in this instance would amount to 315 births in each month.

If you used the device on the basis of necessity, then every birth would be natural *apart from* those which needed induction on medical grounds. In this instance, the total would be the 5% quoted above, totalling 17 or 18 births per month. The overuse of the system on unwitting victims would therefore amount to 1800% more inductions than were necessary, which is ideal for the medicos though not such a self-evidently good idea for the patients.

There are innumerable other examples of the way in which the infectious fashion for believing in computers has come close to harming our lives. The many examples of aircraft near-disasters resulting from this child-like trust in machines that are no less fallible than the people in charge of them are well known. So are the several times when we have come close to war of global dimensions, through the false signalling of an imminent Soviet attack which was due to a computer malfunction.

In the early days of the Vietnam conflict President Lyndon B. Johnson escalated the war against North Vietnam after two ships of the United States fleet were reportedly attacked by submarine missiles in the Gulf of Tonkin. It was only after the Congressional Resolution had been rushed through and the military adventure had increased in scale, that it transpired the reports were based on an error. The sonar records had been confused by echoes from the ships' propellors.

187

In the years of that unhappy conflict there were many later examples in which the computer was used to escalate the war by the mishandling of information and use by trigger-happy war chiefs.

These are not merely the indication that 'things were worse in the pioneer days', either. In the weeks preceding the writing of these words, equally horrendous and costly mistakes continue to occur as they always have done:

+ A bill for a book on computers has arrived in the mail. There is no sign of the book. An enquiry produces the immortal response: DUE TO A COMPUTER FAULT WE ARE UNABLE TO PROCESS YOUR ORDER AT PRESENT.

+ London's first-ever marathon (a 26 mile jog along the lines of the New York race) is marred at the finish as nobody is able to obtain a list of runners. The impressive computer system installed to record the entries as they cross the finishing line becomes hopelessly muddled, and the results are lost.

+ The ultra-sophisticated computerized guidance system of the Polaris submarine *George Washington* allows it to collide with the Japanese freighter *Nissho Maru*, 2,350-tons, or perhaps that should now be 2,387,600 kg, in the East China Sea with the loss of two lives and a great deal of international kudos.

+ The gigantic public-relations exercise behind the launching of *Columbia*, the first space shuttle, is humiliated when the $10,000,000,000 project fails to lift off on time. The delay turns out to be due to an untraceable computer fault. Later, the second launch is postponed when a rogue computer cannot be overruled (and the flight shortened through a major electrical fault).

There are real reasons for us to look closely at the way computers are revered. The old notion, much quoted by computer scientists, was always *garbage in, garbage out* but it is perfectly possible to put in entirely reliable information and still obtain the 'garbage' if the machinery is being controlled by inept operating procedures. These days there is a higher chance of that happening than ever before, as the increasing reverence for computers is being shadowed by falling scholastic standards.

How much personal information is now handled (or mishandled) in these electronic filing-cabinets is unknown. As one example, on the British driving licence is printed the holder's birthdate in one corner. Drivers are told that if they wish to preserve the secrecy about when they were born, to which they may feel entitled, all they have to do is cut off the birthday (there is a special dotted line printed across the corner on which it appears) and the date will be theirs alone.

This is, not to put too fine a point on it, an astonishing falsehood. The date remains on the licence all the time, since it makes up much of the identification number which cannot be removed. The system works like this.

The last two digits of the year are split and put at the front and back of the number respectively; the fourth and fifth numbers are the day, and the second and third numbers the month:

6   07   03   0

decade month day year (i.e. 3 July 1960)

The perfectly deliberate misrepresentation does not matter much, in the sense that coyness about your birthday is hardly a serious subject. What is serious is the way in which a government department can assure you that you do have the right to remove your personal information when all the time they have arranged matters so that you cannot do it at all.

Another example was the introduction of Value Added Tax in 1973. Every company and self-employed individual turning over more than £5000 per annum ($12,000 at the time) had to complete a form giving all the basic trading details of their organization. Among the details required was the bank account number. It was said that the purpose of this was to facilitate the repayment of VAT by the

189

Customs and Excise, should this ever become necessary.

That is patently silly. People are repaid money without any need for them to reveal their personal or business account numbers to some computer overlord. Not only that, but the repayment of money by the tax collectors is very much a minor occupation for them, so there is no excuse for this sudden interest in this matter. However, the forms were a legal requirement, so they were all completed and duly filed by the Customs in their massive computer store at Southend-on-Sea. The result is that every business (apart from those too small to make a living wage) is listed in a computer, along with the access reference to their bank account at any time.

Can you imagine the outcry if the government announced that it wanted a computer file on every business, with their account number at the bank? By doing it subtly, an unprecedented data-bank was established without so much as a murmur of suspicion.

In the United States there is already a vast bank of data on nearly 30,000,000 American citizens. It has been established by the Defense Department, and each day as many as ten thousand new items of data are processed for addition to the store. How the information is presumed to be accurate, when it is collected on such a huge scale, is imponderable.

The question of reliability is a real problem. Since the British version of the zip-code was introduced (the Postal Code everyone is supposed to quote on their letter headings) I have noticed over a dozen versions (I beg your pardon – 1.2 decicodes) appearing on our mail. Because the rest of the address is in plain words as formerly, these incorrectly zip-coded letters arrive just as quickly as the rest of the post, partly because the mechanical handwriting-readers that the GPO have spent so much to perfect do not work and so the letters are read by postmen and come through whether they have the benefits of technology or not.

But if the post code was all we had, and that is the aim of automated letter-sorting, then a wrong code would mean no mail. This is just one indication of the damage that digitilization can do. Once a single figure has been misplaced it is impossible to reinstate it, if all-figure data is all you have to go on.

For example, one of my account cards has my name printed as BRIAN J FOD. Now, that does not cause any great difficulty. Not only is it reasonably obvious that there is a mistake, but it can be seen, and corrected; and if someone tried to look it up in an index the nature of the error would materialize without too much difficulty. But what would you do if you were presented with the code that one newspaper correspondent reported on an insurance premium document?

```
L 5/05236/28481/17663/54714/80522/33393/
51800/00014/05 (repeated twice)
00511/21A4260/12/521H119503/498/2202
13PP/00BB A4/YFZ 57 X98 72XJP 1 FT
1/34982002000014 502 214260 5
```

What if one single figure in that went astray? The profusion of numbers and letters is wildly superfluous! Why, it would generate sufficient alternatives to provide one code each for all the people in the world who have ever lived, or who would live up to the year A.D. 50 thousand million, assuming we were all living nine deep in stacked coffins by then ...

If you make a mistake in a letter (like *Margaret Thatches* or *President Raygun*) it is easy enough to decipher, for letters and words create their own context. The trouble with digital information is that numbers do not mean anything but what they are.

So the invasion of our lives by the cult of the Expert is no small matter. From the way you are born to the manner of your dying, great and earth-shattering decisions are taken behind closed doors by people who may well have closed minds. For the Expert all that matters is that they are seen to have a monopoly on

191

power. Their long and meaningless words, elaborate and expensive apparatus stacked in high-security buildings, the machines that they secretly know no more than anyone else how to use for the good of the people, are all part of today's power structure. The omnipotence of the Expert pronouncement and the unchallengeable right to spend huge sums of money gratifying a lust that evolved only to be gratified; these are the pattern for the future.

No matter how much that looks like pompous tyranny of a peculiarly insidious type, Experts are quick to point out that it is nothing worse than benevolent paternalism. Whatever else you do, they say, do not forget that.

I am trying hard not to, myself.

Index